AGRICULTURE AND THE ENVIRONMENT:
MINERALS, MANURE AND MEASURES

Soil & Environment
Volume 7

Agriculture and the Environment: Minerals, Manure and Measures

by

F. B. de Walle
ENERO-TNO, Delft, The Netherlands

J. Sevenster
Promikron BV, Delft, The Netherlands

KLUWER ACADEMIC PUBLISHERS
DORDRECHT / BOSTON / LONDON

Library of Congress Cataloging-in-Publication data is available.

ISBN 0-7923-4794-3

Published by Kluwer Academic Publishers,
P.O. Box 17, 3300 AA Dordrecht, The Netherlands.

Sold and distributed in North, Central and South America
by Kluwer Academic Publishers,
101 Philip Drive, Norwell, MA 02061, U.S.A.

In all other countries, sold and distributed
by Kluwer Academic Publishers,
P.O. Box 322, 3300 AH Dordrecht, The Netherlands.

Printed on acid-free paper

TABLE OF CONTENTS

ACKNOWLEDGEMENTS

The authors first and foremost wish to thank the Technical Soil Protection Committee (TCB), which advises the Ministers of the Environment and Agriculture in the Netherlands on mineral policies. They commissioned and financed this study and reviewed the text twice. We specifically thank Dr. Joop Vegter and Ir. Koos Verloop for their project management and guidance. We also appreciated the thorough reviews of Dr. Anne De Groot.

A major contribution to this project was made by Ir. Harm Smit of the Ministry of Agriculture, who provided us with relevant materials and data and who was willing to review several drafts. We also thank Drs. Titia van Leeuwen of the Ministry of the Environment for key sources and an extensive review. We also thank their counterparts in other EU countries, Mr. A. Dobbelaere of Belgium, Dr. J.S. Schou of Denmark, Dr. E. Lubbe of Germany, Dr. A.J. Osborne and Mr. R. Unwin of the United Kingdom, for their reviews and commentaries. We are much indebted to Dr. Robert Goodchild of DG IX of the European Commission for his permission to use materials from a study commissioned by the EU. We wish to thank the authors of this study, Dr. Ir. P. Scokart, Ir. P. De Cooman and Dr. M. Donnez.

We also appreciate the contributions of the environmental specialists from many states in the USA and provinces in Canada. Their information provided a broader perspective on the issues involved.

Last but not least, we have been fortunate to enjoy the enthusiastic guidance and support of Ms. Henny Hoogervorst, our editor at Kluwer Academic Publishers. We also thank Mr. Phil Johnstone and the editorial and production staff at Kluwer for their excellent work in the final stage of the production of this book.

The TCB has commissioned this study in order to make a contribution to the public debate on mineral policies in the Netherlands and beyond and would like to invite contributions to this continuing discussion.

Postbox 30947
2500 GX Den Haag
The Netherlands
Tel. 31–70–3393034
Fax: 31–70–3391342

EXECUTIVE SUMMARY

In this study, an overview is presented of agricultural policies on manure and minerals, relating to the Nitrate Directive to remedy excessive surface- and groundwater contamination from intensive agricultural practices. Six countries belonging to the European Union were studied: the Netherlands, Belgium, Denmark, France, Germany and the United Kingdom. The policies and their legal incorporation were related to agricultural and environmental conditions in each country. In addition, an inventory was made of agricultural mineral policies in the United States and Canada. Conditions for livestock farming in North America differ considerably from those in Europe, but their solutions shed a different light on European policies.

Research has shown that there are still very considerable mineral surpluses in many countries and regions. In both the Netherlands and in the Flemish part of Belgium, existing problems due to very high levels of manure production are structural rather than local and cannot easily be solved by transport of manure to other regions. To a lesser extent, Germany, Denmark and relatively small parts of France (Brittany) and the United Kingdom, still exceed the norms for an equilibrium fertilization. In Denmark, existing problems can probably be solved within the existing legislative framework. The Netherlands, Flanders, several German Länder (Nordrhein-Westfalen and Schleswig-Holstein) and Brittany, because of the intensity of their livestock sector, will require more structural solutions. In these countries, nutrient surpluses have stabilized and some moderate reductions have been achieved since 1985. The Action Programs comprise new measures to realize further reductions, but in most cases the political will to enforce stringent regulations and impose significant levies is not there. Thus, it remains to be seen whether the Action Programs will be able sufficiently to reduce the present surpluses or whether more structural solutions need to be found.

In the United States, agricultural mineral policies focus mainly on large feedlot operations. Treatment requirements for feedlots are comparable to discharge standards for industrial waste waters. In Canada, several provinces use voluntary regulations to gain economic benefits and volunteer inspectors, as a means of using social control to enforce compliance with environmental norms. Violators of environmental regulations are fined quite heavily.

Based on the results of this study, the following recommendations are made. In those countries where agricultural mineral surpluses are structural, a choice will have to be made between a drastic reduction of the intensive livestock sector or the development of other alternatives. A possible alternative is suggested by policies in the USA, where large livestock operations are bound by treatment requirements for waste water that are the same as for industrial waste. On the basis of this model, a policy approach is recommended that differentiates between large, intensive, industrial type livestock operations and small, family-type, non-intensive livestock farms. The large operations, that are no longer dependent on land area, would be subject to industrial waste treatment norms. The economies of scale that can be achieved because of their size should enable them to meet the necessary treatment costs. The small family farms should have a sufficiently large land area, or contracts with other farms, to dispose of their manure. A specialization as ecological or biological farm would help them find a market niche.

CHAPTER 1: INTRODUCTION

INTRODUCTION

The Dutch Technical Soil Protection Commission on Soil Protection (TCB) is conducting an international study of the mineral policies in the agricultural sector, especially with respect to animal manure and land fertilization, in European countries and American States. The different approaches will be evaluated, when and if this is possible, as to their effectiveness and will be compared to the Dutch policy.

Deterioration of the environment as a result of the emission of minerals from agriculture has created increasingly serious problems during the last decades. In the Netherlands, phosphorus incorporation in the soil has exceeded its maximum capacity in many areas, causing direct releases to ground and surface waters. Numerous drinking water wells have exceeded WHO norms. Rising nitrate levels in the aquatic environment lead to the eutrophication of surface and marine waters, while contamination of drinking water, due to nitrates in groundwater, pose serious risks to human health. Drinking water standards of 50 mg/l nitrate allowable in drinking water have been agreed upon by European legislation (EU drinking water standards).

As a result of high cattle and pig densities, excess animal manure is polluting ground waters (exceeding the drinking water criteria) and causing eutrophication of lakes; rivers carry nutrients to sea, contributing to its eutrophication risk. Nitrate pollution has become a serious political issue in the European Union. Thus, a directive concerning the protection of waters against nitrate pollution from agricultural sources was announced to Member States in 1991. One of the main elements of the Nitrate Directive is that the application of animal manure should not exceed 170 kg of nitrogen per hectare in designated zones, which are vulnerable to the leaching of nitrates.

An assessment of the number of agricultural holdings in the EU which may be affected by the standard of 170 kg N/ha, shows that almost 600 000 holdings, or 13% of all holdings, might be affected (Brouwer *et al.* 1996). The production of animal manure on these holdings with excess animal manure is on the average 350 kg N/ha. The proportion of holdings with excess animal manure is lowest in France (6%), Ireland (8%) and Italy (6%) and highest in Belgium (47%) and 63% in the Netherlands (Frederiksen, 1996).

Most countries in the EU agree that high emission levels of minerals from modern agriculture need to be regulated. The Dutch Ministry of Agriculture allocated 235 million ECU in a restructuring fund to purchase manure production rights during the coming 7 years. A 65 million ECU incentive program stimulates the development of emission reducing barns, covering of manure silos and advanced manure land applicators to reduce ammonia losses. Similar initiatives have been taken by governments of other EU countries, even though their policies may show substantial differences in interpretation of the Nitrate Directive, which may be related to differences in the extent of the problem for each country.

OBJECTIVE OF THE PROJECT

It is the objective of this report to provide an overview of the agricultural policies on minerals and manure of different states with respect to:

- environmental pressures and sources for each country;
- ambient conditions of soils, surface-, ground- and drinking water;
- need and purpose of the policies and regulations;
- policy formulation and legal incorporation;
- implementation of policies at the national and local level;
- effect of measures;
- future developments.

Three aspects are especially important, i.e. the goals of the policies, the legal incorporation and the formulation and implementation. These are related to the types of agricultural practices and the current and desired environmental quality. An overview will be made of the approaches in different European countries, American States and Canadian provinces, realizing that in North America conditions for livestock farming are quite different from those in Europe. Nevertheless, their solutions may shed a new and different light on European policies.

The second aim of the current project is to use the overview to make a comparison of the different policy approaches in selected countries and to provide an interpretation of the observed differences.

Third, we hope to be able to draw some conclusions and make recommendations in order to assist in and make a contribution to the intense public discussion in this area.

PROJECT APPROACH

The project was conducted in two phases. The first phase covered the general characteristics of the mineral policies and regulations in a large number of countries. It addressed the different policy aspects (character, instruments, implementation), national mineral budgets (input, output, crop uptake), livestock characteristics (composition, size, expansion), application of the manure to agricultural soils in relation to other fertilizers, quality of the environment as threatened by mineral excess. The main criterion for collecting the material was its availability and ease of collection.

In the second phase, a selected number of countries were studied in greater detail and additional materials were collected. Six countries were selected that belong to the European Union (Netherlands, Belgium, Denmark, France, Germany and the United Kingdom) and the United States and Canada. Several interviews were conducted to obtain detailed backgrounds. The collected materials were placed in a common framework. The detailed information of the second phase allows an in-depth comparison of the different approaches and its interpretation.

The following questions had to be answered:

Environmental quality questions
- What are the environmental pressures and sources in each country?
- What are the ambient conditions of soils, surface-, ground- and drinking water?
- What are the desired quality goals for the receiving waters?
- What is the relation between eutrophication and N and P content of these waters and what is the manure contribution?

Policy questions
- What is the purpose of the existing policy?
- What is the legal incorporation?
- What are the regulatory controls that were chosen and on what basis?
- What are the regulations for manure application to the land in general and in selected areas (watersheds for drinking water, nature areas)?

Regulatory questions
- Are the regulations formulated for one or more minerals (N, P, K) in manure and/or mineral fertiliser?
- What norms are used for the application of manure and/or mineral fertiliser: generic application norms or other norms, based on individual circumstances?
- Which countries use a mineral accounting system, to realize or approximate equilibrium applications in which the mineral application equals crop uptake and removal and losses without gradually accumulating in the soil?
- Which countries have areas, where intensive agriculture and nature preserves coexist?
- What are the regulations for record keeping and verification and what are the financial measures and incentives used?

Effects and future developments
- Is it possible to ascertain the effects of the measures taken so far?
- What are the policy goals for the future and what future developments does each country envisage?

SOURCES OF INFORMATION

One of the problems inherent in collecting data about different countries, in such a manner that international comparisons can be made, is the comparability of the data collected. If the information one collects is based on different assumptions for different countries or has used a different methodology, it becomes very difficult to make comparisons and draw conclusions. In collecting the necessary information for this study, we tried in the first place to find data and research studies that apply to all countries studied, at least all the European countries. Fortunately, there were some excellent studies of manure and fertilizer use in all EU countries, both at a regional level and at farm level and Parcom data on the effects of measures.

One important source was the study *Les Codes de Bonne Pratique en relation avec la Directive Nitrate* by de Cooman, Scokart and Donnez (1995), that was commissioned by the European Commission. This study has collected data on manure production and the use of inorganic fertilizer and also on ambient conditions in all EU countries. The study attempts to predict the potential impact of the Codes of Good Agricultural Practice on groundwater and surface water quality in the EU countries involved. The report pays little attention to the different agricultural mineral policies used by the different countries. The present study will utilize a lot of the information presented in the EU study, but will expand on it by focusing on the differences in the agricultural mineral policies of the EU countries and their effectiveness. It will also include chapters on mineral policies in the United States and Canada, which may shed a new light on European policies.

Some studies of agricultural mineral policies across the European Union were also available. Other sources applied only to one country at the time, making data hard to compare to those of other countries. In the USA and Canada, agricultural mineral policies are less of a concern than in Europe. Both the USA and Canada have some federal policies concerning feedlots, and some states and provinces have their own policies and regulations. To make an inventory of these state and provincial policies, a letter and a questionnaire were sent to all states and provinces, asking for answers to specific questions and for more information, if available. The questionnaire data provided some data that are comparable across those states and provinces that answered.

METHODOLOGY

In the search for information, it became apparent that the study commissioned by the EC relating to Codes of Good Agricultural Practice in relation to the Nitrate Directive (de Cooman *et al.*, 1995) provided information that was very pertinent to this project. However, some of the data it provides require an explanation of the methodology used.

One objective of the study was to make an assessment of the extent of the manure problem in different EU countries and regions thereof. To assess the manure surplus by NUTSII region (regions selected by the Working Group on Nutrients), they calculated the following quantities:

- **Manure production.** The report provides data on manure production by region, based on Eurostat statistics about the structure of the livestock in that region and the norms used for manure production of each kind of animal. These norms are presented in Table 1.1. Conversion coefficients to Livestock Units

(LU) are given in Table 1.2. Transfers of manure between regions have not been taken into account.

- **Utilization of mineral fertilizer.** Estimates for the use of mineral fertilizer by region have been made on the basis of data supplied by the member states. These data are often supplied by manufacturers and suppliers and estimates are made for each region.
- **Calculation of total fertilization.** The total fertilization is the sum of manure production and the utilization of mineral fertilizer. Volatilization of ammonia has not been taken into account; the assumption is that the manure will be homogeneously divided across different crops.
- **Limits.** For each type of fertilizer (manure, mineral and total), fertilization

Table 1.1   Norms for the production of nitrogen from different animals in 5 countries in kg N/animal/year

Animal: variable in REGIO database	Belgium (1)	Germany (2)	Denmark (3)	France (4)	Holland (5)
BOVIN2	10.90	15.00	38.30	17.30	25.40
BOVIN3	33.48	15.00	38.30	17.30	25.40
BOVIN4	33.48	15.00	38.30	17.30	25.40
BOVIN5	55.80	50.00	43.80	50.40	78.00
BOVIN6	55.80	50.00	38.30	50.40	78.00
BOVIN7	55.80	50.00	38.30	50.40	78.00
BULLS	79.30	85.00	43.80	59.10	100.00
BOVIN9	79.30	85.00	38.30	59.10	100.00
HEIFERS	79.30	85.00	38.30	67.20	100.00
DAIRY COW	87.15	85.00	126.30	84.00	144.00
OTHER COW	87.15	85.00	73.30	58.70	120.00
BUFFALOS	61.00	61.00	61.00	61.00	61.00
PIGLET1	3.20	3.20	3.20	3.20	3.20
PIGLET2	9.91	11.00	14.50	10.10	13.90
PIGLET3	9.91	11.00	14.50	10.10	13.90
VERRAT	16.75	18.30	22.90	20.10	17.20
SOW	16.75	29.00	30.90	39.70	35.00
SHEEP	10.46	7.00	15.00	8.10	21.80
GOAT	10.46	7.00	15.00	8.10	15.00
EQUID	87.15	68.60	50.00	67.20	45.00
POULTRY	0.50	0.50	0.70	0.60	0.60

The norms for Belgium have also been used for the UK

(1) Save "Mestdekreet" (1991)

(2) Faustzahlen (1993) – SCHLEEF et KLEINHANSS (1994)

(3) RUDE (1993)

(4) RAINELLI (1993) in SCHLEEF et KLENHANSS (1994)

(5) BROUWER (1993) in SCHLEEF et KLEINHANSS (1994)

Source: de Cooman *et al.*, 1995

Table 1.2 Conversion Coefficients to Livestock Units (LU)

Animal: variable in REGIO database	LU
BOVIN Cattle2	0.400
BOVIN Cattle3	0.400
BOVIN Cattle4	0.400
BOVIN Cattle5	0.600
BOVIN Cattle6	0.600
BOVIN Cattle7	0.600
BULLS	1.000
BOVIN Cattle9	0.800
HEIFERS	0.800
DAIRYCOW	1.000
OTHERCOW	1.000
BUFFALOS	1.000
PIGLET1	0.027
PIGLET2	0.300
PIGLET3	0.300
VERRAT	0.300
SOW	0.500
SHEEP	0.100
GOAT	0.100
HORSES	0.600
POULTRY	0.012

limits have been calculated which are based on the norms contained in the Code of Good Agricultural Practice in each country. The norms used were those utilized at the time the study was conducted, and may have been changed since then. These norms often constitute a compromise between what is desirable for agriculture and what is acceptable for the environment.

- **Difference** between the limits and the actual fertilization is calculated for each region. The differences that are thus calculated provide a rough measure of the agricultural mineral surplus by region and give an indication of how structural the mineral surplus problem in that region, and consequently in that country, is.

Another objective was to calculate the theoretical impact which the measures A1–A5 and A6 (see Chapter 2) of the Codes of Good Agricultural Practice could have on the quality of groundwater and surfacewater, if they were fully implemented. To this end, the following quantities have been calculated:

- Direct losses due to the use of mineral fertilizer;
- Direct losses due to manure on grassland;

- Direct losses due to manure in stables;
- Direct losses due to leaching of silos or storage of manure;
- Losses due to run-off.

With these quantities, calculations have been made according to a scenario, where these losses are reduced in line with the measures adopted in the Code of Good Agricultural Practice, to predict the theoretical impact of these measures in each region.

Three other theoretical effects were calculated on the basis of three scenarios, that were based on three different assumptions: first, the norms for manure application are respected; second, the norms for mineral fertilizer are respected; and third, the norms for both manure and mineral fertilizer are respected (equilibrium fertilization). For each scenario, the theoretical effects on surface water, shallow groundwater and deep groundwater were calculated, with fixed percentages for denitrification: 10% for surface water, 40% for shallow groundwater and 70% for deep surface water. The theoretical effects are a measure of the reductions in agricultural minerals that still have to be realized for a particular region or country. In this report, these limits and theoretical effects will be referred to for each country.

The methodology used in the study by Scheef and Kleinhanss (1996), who have calculated regional mineral balances for the EU countries, also requires an explanation. The mineral balance equals input (nitrogen from mineral and organic fertilizer, plus deposition) minus output (nitrogen uptake by crops). The figure for organic fertilizer, Table 1.3, still has to be corrected for ammonia losses (30%) to arrive at the correct balance figure.

THE STRUCTURE OF THIS BOOK

This report will be structured as follows:

Chapter 2. The European Context
Since all the present legislation at a national level is taking place in the context of the European Union, which in itself is also in the process of evolving, it is important to provide some information about some of the earlier developments at the European level that have played an important role in the process. The International North Sea Conference (INSC), the Paris and Oslo Conventions (PARCOM, OSPAR) and the European Community (EC) have all made contributions to the process. This will be reviewed in the next chapter.

Chapter 3–10. The individual countries
For each country, an overview will be presented of the structure of agriculture, environmental pressures (manure production and use of mineral fertiliser) and ambient conditions of groundwater and surface water), agricultural mineral policies, regulations, financial incentives and future developments. The information available for the USA and Canada is more limited, so not all topics may be covered;

Table 1.3 Nitrogen balances for EU countries and regions with surpluses greater than 100 kg/ha

Region	Depositions (kg/ha)	Fertilizer (kg/ha)	Manure (kg/ha)	Uptake (kg/ha)	Balance (kg/ha)
EU-11	16	83	65	79	65
Germany	30	129	82	104	113
Schleswig-Holstein	26	158	92	129	120
Braunschweig	35	175	37	117	119
Hannover	35	153	71	114	124
Lüneburg	35	122	85	102	114
Weser-Ems	35	124	146	106	155
Bremen	37	97	95	94	107
Düsseldorf	38	151	113	122	146
Köln	38	159	71	123	125
Münster	38	128	151	104	167
Detmold	38	134	102	106	138
Arnsberg	38	128	89	106	122
Darmstadt	30	131	63	102	103
Gießen	30	133	71	107	105
Kassel	30	143	71	112	111
Koblenz	26	136	67	107	102
Trier	26	133	89	114	108
Stuttgart	25	137	87	112	111
Tübingen	25	132	96	113	111
Oberbayern	28	134	100	117	116
Niederbayern	28	137	91	114	116
Oberpflaz	28	138	89	115	114
Oberfranken	28	133	73	107	105
Mittelfranken	28	140	95	116	119
Unterfranken	28	150	52	106	109
Schwaben	28	147	118	129	129
Sachsen	31	143	83	118	114
Thüringen	30	119	72	99	101
France	17	91	52	81	63
Bretagne	17	108	147	98	130
Italy	11	44	54	65	28
Lombardia	23	87	149	108	107
The Netherlands	36	218	339	171	321
Groningen	36	221	146	151	207
Friesland	36	283	316	187	353
Drenthe	36	189	201	147	219
Overijssel	36	227	446	188	387
Gelderland	36	219	513	182	432
Flevoland	36	193	75	181	100
Noord-Brabant	36	199	587	156	489
Limburg	36	201	479	153	419

Table 1.3 *Continued*

Region	Depositions (kg/ha)	Fertilizer (kg/ha)	Manure (kg/ha)	Uptake (kg/ha)	Balance (kg/ha)
Utrecht	36	259	434	202	396
Noord-Holland	36	218	207	169	230
Zuid-Holland	36	204	225	186	211
Zeeland	36	189	63	150	119
Belgium	33	161	208	162	178
Brabant	33	176	119	139	153
Antwerpen	33	203	453	196	358
Limburg	33	175	220	161	201
Oost-Vlaanderen	33	203	320	174	286
West-Vlaanderen	33	199	353	171	308
Hainaut	33	162	128	148	136
Liege	33	151	150	171	119
Luxembourg	27	128	118	119	119
United Kingdom	18	107	66	91	80
Humberside	26	174	43	121	109
Leices., Northham.	28	148	61	115	103
Lincolnshire	28	176	27	116	107
Cheshire	26	137	109	123	117
Lancashire	26	101	106	96	105
Ireland	10	60	68	71	47
Denmark	18	141	102	127	104
Vest for Storebaelt	18	141	116	127	113
Spain	6	35	39	46	22
Portugal	4	32	38	44	18

Source: Schleef and Kleinhanss, 1996

also, only some states and provinces have legislation dealing with agricultural mineral policies.

Chapter 11. International Comparison
A comparison will be made between the policies of the different countries and, where possible, their effectiveness. In this process, we will take into account the actual conditions of each country concerning the structure of its agriculture, ambient conditions of groundwater and surface water and economic conditions.

Chapter 12. Conclusions and Recommendations
On the basis of the information presented, conclusions will be drawn and some recommendations made.

CHAPTER 2: INTERNATIONAL POLICIES

INTRODUCTION

In this chapter, international conferences and commissions that have dealt with nutrient pollution of groundwater and surface water, will be reviewed. These international policies have created the context in which national policies have been developed. International treaties frequently give an important impetus to policies on a national level (Vosmer, 1995).

THE FIRST INTERNATIONAL NORTH SEA CONFERENCE (INSC)

Participants were the ministers of all North Sea States: Belgium, Denmark, Germany, France, The Netherlands, Norway, Sweden and the United Kingdom. A Commission of the European Community also participated. Observers from other states were also present.

Preparation
In the preparatory phase, a meeting took place in Bonn where a German report on water quality of the North Sea was discussed. In this report, the opinion was expressed that regional discharges should not be allowed to affect the ecosystem of the North Sea. Also, it would be better not to wait for further scientific research, because action was needed to prevent extensive damage. Participants reached an agreement on the need for preventive measures, which became the starting point of the first INSC, which took place in Bremen in 1984. The objective of the conference was:

- To provide political impetus by agreements at the ministerial level.
- To facilitate the efficient implementation of existing international rules by the competent authorities.

Decisions
To reduce the nutrient supply from agriculture, the following action agreements were formulated:

- The capacity of manure storage facilities should be sufficient to allow application of manure only at those times that allow uptake by the crops.
- The improvement of agricultural practices for the optimal use of nutrients and prevention of the application of manure and mineral fertilizer at inappropriate times.
- Correct management of livestock.

11

- Implementation of protection areas for surface waters.
- Facilitating research, monitoring and information about the dispersion of nutrients in the environment, so that farmers can be informed.

THE SECOND INSC

The second INSC took place in London in 1987. Its objectives were the same as the first conference, plus the approval of the report on the implementation of measures that had been agreed upon during the first INSC and discussion of further measures as they were needed. The main decisions were:

- A 50% reduction of the nutrient supply to the North Sea, both from point sources and from non-point sources.
- Taking efficient measures at the national level to decrease nutrient emissions to parts of the environment where they may do damage.

THE THIRD INSC

The third INSC conference took place in the Hague in 1990. The objectives were:

- Evaluation of measures taken so far.
- Implementation of measures to reduce the supply of noxious substances by 50%.
- To end discharges and burning of industrial-chemical wastes in and at sea.
- Protection of the Waddenzee.
- To research the possibility for further cooperation and the possible role of the Paris Commission in it.

Decisions that were made were:

- Designation of coastal zones, including the Skagerrak, as problem areas with regard to eutrophication. In case of presence or supply of increased concentrations of nutrients, areas can be labeled as potential vulnerable zones.
- More measures are needed to achieve the 50% reduction of nutrients before 1995.
- Further measures for agriculture have been agreed upon.

THE FOURTH INSC

The fourth INSC took place in 1995. A report was presented on the progress made so far. It has become clear that the objective of a 50% reduction has not been realized, especially not for nitrogen.

Decisions that were made were:

- The goal of 50% reduction of nutrients to the North Sea should still be realized.

- All countries that border on the North Sea should designate their whole territory as Vulnerable Zone (Nitrate Directive). This applies only to a part of France.
- More measures should be taken to stimulate ecological agriculture.
- Measures should be taken to ensure the set-aside of 10% of the agricultural area.

THE PARIS CONVENTIONS AND OSLO COMMISSION (PARCOM, OSCOM AND OSPAR)

Members of the Paris Convention are: Belgium, Denmark, Germany, France, Netherlands, Norway, Sweden and the UK, Ireland, Portugal, Iceland and the EEC. There are also several observer countries, international organizations and environmental organizations present.

Objectives and programmes

The members are required to do everything possible to prevent the pollution of the sea from sources on land and in the air, like industry, agriculture, deposition and communal sources. Measures should be taken to prevent this pollution.

Programs for nutrient reduction have been formulated during several Paris Conventions:

PARCOM Recommendation 88/2

Parties to the treaty are required to aim for a considerable (50%) reduction of nutrient supply to the North Sea between 1985 and 1995, mostly the supply of nitrates and phosphates to areas where they can do damage.

PARCOM Recommendation 89/4

These recommendations are a coordinated programme to reduce the supply of nutrients.

PARCOM Recommendation 92/7

Two agreements were made. The first deals with nutrient supplies in four categories:

- ammonia emissions;
- leaching of N, mostly nitrate;
- leaching, run-off and erosion losses of phosphorus;
- agricultural emissions.

The second agreement requires all parties to implement all or part of those measures, which are either mandatory or advisory in nature.

PARCOM Mineral balance

Calculating the mineral balance of the national agriculture gives insight into nutrient surpluses. These surpluses comprise all losses to the environment. Changes in the mineral balance give an indication which measures have been successful.

By making it mandatory to keep a mineral balance in agriculture, it becomes clear where nutrients are being wasted, and this can lead to better measures.

PARCOM: Working Group on Nutrients (NUT)
Objectives of NUT are:

- Defining quality objectives for the area to counteract supplies of nutrients, collecting data and formulating methods to achieve these quality objectives.
- Information exchange about implemented and proposed measures against the pollution by nutrients of riverbeds, groundwater, coastal waters and lakes.
- Evaluation of the potential and actual effects of the reduction of nutrients by coordinated reduction programmes.
- Research on the effects of nutrient pollution and the methods of reducing the effects of the pollution by nutrients, possibly in coordinated programmes.

PARCOM: Working Group on Nutrients and Agriculture (NUTAG)
Within NUT, a special working group is concerned with nutrient emission from agricultural sources.

OSCOM and OSPAR
More or less parallel to PARCOM, which has focused on nutrient pollution from sources on land and in the air, OSCOM has focused on nutrients from sources on sea, like shipping, off-shore drilling and dredging. In OSCOM, all PARCOM countries participate, plus Ireland, Spain, Portugal, Iceland and Finland.

In 1992, a proposal was accepted to combine PARCOM and OSCOM into OSPAR. The ratification of this decision by member countries will be completed in 1997. OSPAR has two objectives, which are monitoring and assessment (ASMO committee) and taking measures for the protection of the sea (PRAM committee, Programs and Measures).

THE NITRATE DIRECTIVE OF THE EUROPEAN UNION

Background
The origin of the EU Nitrate Directive goes back to the time before the Maastricht Treaty of 1992, when the EU was still the EEC (European Economic Community). The environmental objectives of the EEC were, since the European Act of 1987, where environmental concerns were included in the European Treaty:

- To maintain, protect and improve the quality of the environment.
- To contribute to the protection of human health.
- To ensure the careful and rational management and use of natural resources.

The Nitrate Directive
The objective of the Nitrate Directive is the protection of waters against the supply of nutrients and it requires the member states to take measures to reduce the

supply of nutrients to groundwaters and surface waters. These measures have to be taken within a certain time frame and consist of the following provisions.

Before December 1993 all waters must be monitored and member states have to designate zones, that are vulnerable to nitrate leaching. These Vulnerable Zones are defined as land areas where agricultural production contributes to drinking water problems (the limit of 50 mg nitrate/l) or to the eutrophication of aquifers. The regulations laid down in the Directive are compulsory only in these zones.

A Code of Good Agricultural Practice has to be formulated, which specifies measures that can be implemented on a voluntary basis to improve agricultural practice. This Code should contain the following elements:

- A1. Periods when fertilizers should not be applied to agricultural land.
- A2. Application of fertilizers on steeply sloping ground.
- A3. Application of fertilizers on waterlogged, inundated, frozen or snow covered soil.
- A4. Application of fertilizers near surface water.
- A5. Storage capacity and construction requirements for manure storage facilities.
- A6. Application methods for manure and fertilizer on agricultural land to maintain nutrient losses at acceptable levels.

Four more elements may be included in the Code, but their inclusion is optional:

- B7. Land use management, including crop rotation systems.
- B8. Use of catch crops (crops planted for the purpose of taking up surplus nitrogen).
- B9. Establishment of fertilizer plans on a farm-by-farm basis and keeping records of fertilizer use.
- B10. Irrigation system management.

Furthermore, member states should formulate Action Programs before December 1995, for all Vulnerable Zones, that are mandatory, which need to include the following subjects:

- storage capacity and manure storage facilities; this capacity must exceed that required for storage throughout the longest period during which land application in the Vulnerable Zones is prohibited, with possible exceptions;
- periods when fertilizers should not be applied to agricultural land;
- limitations to the application of fertilizers and agricultural lands depending on: soil condition, type of soil, slope, climate, precipitation and irrigation, land use and exploitation and crop rotation.

These programmes have to be implemented before December 1999. The following limits on the application of fertilizers should be implemented: 170 kg N/ha by 2003. In the period until 1999, the limit is 210 kg N/ha. Compliance with these standards is most crucial in areas with intensive animal husbandry and most difficult to achieve. The Directive allows for use of other standards, but these must be based on objective criteria, and should not violate the objectives of the Directive. Where the Action Programme is deemed to be insufficiently based on monitoring data, further action has to be taken. A reviewing process must be organized once every 4 years.

Member states are also required to implement monitoring programs to monitor the nitrate concentrations of ground- and surface waters. The implementation status of the Nitrate Directive in the EU countries is shown in Table 2.1.

LEGAL STATUS OF DECLARATIONS

Legal status of INSC declarations
The INSC declarations are non-binding decisions. They must be seen as international policy programs of measures, that can become binding at a later stage, when they become incorporated into law at an international, European or national level. Parties to these agreements are politically committed to them: Ministers declare themselves to be responsible for the protection of the North Sea and non-compliance will imply international loss of face.

The legal status of the PARCOM declarations
The decisions are binding for the parties involved. The parties to the treaty are required to implement those decisions that apply to them. This does not mean they have to incorporate these decisions into national laws or regulations. The recommendations and agreements are non-binding. OSPAR decisions are less binding than EU directives.

The legal status of EU declarations
Independent of the kind of declarations there is the possibility that they are in conflict with the national legislation. In that case national law has to give precedence to European law.

Regulations
Regulations are binding in all parts and apply directly to all member states. This means that regulations are legal and do not have to be implemented any further.

Directives
Directives are binding for all member states with respect to the stated objectives. The implementation of directives is left to the member states. National law needs to be adjusted within the given period. The European Court requires that directives, in combination with national law, are transposed into national law.

Decisions
Decisions are binding in all respects for member states that are signatories to them.

Table 2.1 Implementation status of the Nitrate Directive as of 30–7–1997 in EU countries: Transpostion into national law and Codes of Good Agricultural Practice, Designation of Vulnerable Zones and Action Programmes

Country	Transposition of the Directive into National Law		Codes of Good Agricultural Practice	Designation of Vulnerable Zones		Action Programmes	
	Date of communication	Conformity of measures	Date of communication	Date designations completed	Area covered	Date of notification	In compliance
Belgium	–	–	–	–	–	–	–
Denmark	17–12–1993	Yes	17–12–1993	12–07–1993	Whole territory	08–01–1996	Under examination
France	27–08–1993	Yes	10–02–1994	–	46% of agricul. land*	–	–
Germany	01–04–1996	No	01–04–1996	07–11–1994	Whole territory	01–04–1996	No
Netherlands	–	–	05–01–1994	05–01–1994	Whole territory	–	–
United Kingdom	28–06–1996	No	10–11–1994	10–02–1997	69 Vulerable Zones*	–	Under examination
Austria	26–01–1996	Check ongoing	26–01–1996	26–01–1996	Whole territory	11–11–1996	Under examination
Finland	24–03–1995	No	24–03–1995	–	–	–	–
Greece	–	–	05–05–1994	–	–	–	–
Ireland	17–07–1995	No	20–08–1996	17–07–1995	No Zones*	–	–
Italy	–	–	22–12–1993	–	–	–	–
Luxembourg	26–10–1994	Yes	25–03–1996	19–10–1994	Whole territory	19–10–1994	No
Portugal	–	–	–	–	–	–	–
Spain	11–03–1996	Yes	–	–	–	–	–
Sweden	25–01–1996	Check ongoing	25–01–1994	25–01–1996	5 Vulnerable Zones*	25–01–1996	Under examination

*These decisions are currently being examined by the Commission

Source: EU 1997

THE NETHERLANDS

CHAPTER 3: THE NETHERLANDS

COUNTRY CONDITION

The agricultural sector in the Netherlands occupies about 2 million ha, which means that half of the total territory of the country (about 4 million ha) is dedicated to agriculture. It involves 110 000 farms, which employ 286 000 people. Most of these farms are small family enterprises: the average size of general cropping farms is about 30 ha, of dairy farms 20 ha, while horticultural farms are generally no bigger than 1 ha. Granivore farms, mostly pig or poultry rearing operations, generally also cover a small area, even though their operations may be very intensive (de Cooman *et al.*, 1995).

Even though the area devoted to agriculture is not that large, the economic importance is significant. The Netherlands is not only one of the smallest, most densely populated, highly industrialized countries in the world, it is also one of the larger net exporters of agricultural products worldwide (Baldock and Bennett, 1991). In 1992, 25% of all exports were agricultural products. This is possible because the production is very intensive. Many millions of ECU have been invested in buildings, mechanization and automization, fertilizer use is extensive. The combination of considerable research efforts with a well-developed agricultural extension service has always been geared towards improving productivity, even though this is now balanced by environmental concerns. The total value of agricultural production amounts to 15 700 million ECU and constitutes 5% of the GNP.

Farms and related businesses form an integrated system called the 'agrobusiness complex': it consists of cooperatives and vertical associations that involve production, trade and the food industry. In horticulture, 'veilingen' (auctions) enable the rapid confrontation between supply and demand.

The Netherlands have at their disposal about 8 million livestock units (LUs) or 8% of the European livestock population. This population consists of 49% cattle, 37% pigs and 12% poultry. The total livestock population has not changed much between 1983 and 1989, but the reduction in cattle has been compensated by an increase in the number of pigs. Since 1990, the increase in the number of pigs has been stabilized. The country has been more than self-sufficient for sugar, poultry, pork, cheese, eggs and butter, with self-sufficiency rates varying between 200 and 400%. These rates suggest that livestock farming is crucial to the Dutch agricultural economy: in 1992, it was responsible for 60% of the total agricultural production value. However, the high intensity of livestock farming can only be maintained by importing large quantities of animal feed, with the result that the livestock economy generates a substantial net surplus of nutrients. The most serious manifestation of this surplus is the large quantity of surplus manure, which cannot be disposed of through normal agricultural practices.

Within the Netherlands, livestock operations are concentrated on the sandy soils of the following provinces: Overijssel, Gelderland, Limburg and Noord-Brabant together possess about 70% of the national livestock population. Livestock density is highest in these areas and varies from 5 LU/ha in Overijssel to 8 LU/ha in Noord Brabant. Eight of the total of 12 Dutch provinces have livestock densities over 2 LU/ha: only Zeeland, Noord-Holland, Flevoland en Groningen have lower densities. Further analysis shows that cattle densities do not vary much between the different provinces (between 2 and 3.5 LU/ha), but pig and poultry densities do. Most granivore farms are located in the sandy regions in the east, middle and south and here densities can reach 8 LU/ha. Because livestock densities are so high in these regions, this is where most of the manure surplus manifests itself (de Cooman *et al.*, 1995).

ENVIRONMENTAL PRESSURES AND SOURCES

In this section, we will make an inventory of the different sources of N and P that contribute to elevated N and P levels of surface- and groundwater. In the Netherlands, concern about phosphate has been relatively high, compared to other countries, because phosphate-saturated soils are a common problem in the sandy regions in the south and east of the country. In the EU, the main emphasis is placed on controlling nitrogen sources to protect drinking water quality. In the Netherlands both elements are creating environmental pressures.

Use of mineral fertilizer
Utilization of mineral fertilizer in the Netherlands is very high compared to other European countries, particularly in grassland areas and greenhouse horticulture fertilizer (315 versus 835 kg N/ha). Fertilizer use in the Netherlands increased from 405 000 tons N in 1970 to 504 000 tons in 1986, but decreased to 380 000 tons nitrogen in 1992. Average use of mineral fertilizer is 218 kg N/ha for the whole country (see Table 1.1) and varies from 283 kg N/ha in Friesland to 189 kg/ha in Drente (Table 3.1).

Use of animal manure
Even though levels of mineral fertilizer use are high in the Netherlands, it also produces and utilizes considerable quantities of animal manure. The density of the livestock population and thus the production of nitrogen in organic form in livestock farming is at the root of the high level of manure utilization in arable farms and horticulture, on grasslands and maize. Other forms of organic nitrogen such as compost and sludge, are also used.

For the country as a whole, the average utilization of manure is 339 kg N/ha. The highest level of manure supply is found in Noord-Brabant, where it amounts

Table 3.1 Mineral fertilizer (kg N/ha per year)

NUTS II	Limit	Utilization	Difference
Utrecht	385.6	258.9	−126.7
Overijssel	350.5	227.0	−123.5
Gelderland	339.8	219.0	−120.8
Zuid-Holland	317.8	204.0	−113.8
Friesland	387.7	283.0	−104.7
Noord-Holland	322.1	218.1	−104.0
Drenthe	285.2	189.0	−96.2
Noord-Brabant	282.1	198.9	−83.2
Limburg	258.0	201.3	−56.7
Groningen	265.1	220.9	−44.2
Zeeland	210.0	189.0	−21.0
Flevoland	203.0	193.1	−9.9

Source: de Cooman *et al.*, 1995

to 587 kg N/ha, but not all of it is applied locally: some is transported to other parts of the country to be utilized mainly in arable farms. Zeeland has the lowest level of manure production, 63 kg N/ha (Table 3.2). Research on the average supply of manure at farm level shows, that in the Netherlands 63% of all holdings (cropping and livestock farms) produce more than 170 kg N/ha, the EU standard, and 99% of all farms producing manure produce more than this level (Table 3.2). A further analysis of all farms with manure production exceeding

Table 3.2 Organic fertilizer (kg N/ha per year)

NUTS II	Limit	Production	Difference
Zeeland	179.4	64.2	−115.1
Flevoland	177.4	76.5	−100.9
Groningen	199.3	152.9	−46.4
Noord-Holland	210.4	212.8	2.4
Drenthe	204.0	210.0	6.0
Zuid-Holland	209.1	229.1	20.1
Friesland	231.3	328.3	97.1
Utrecht	226.5	434.2	207.7
Overijssel	223.2	473.2	250.0
Limburg	193.8	481.6	287.8
Gelderland	216.2	529.0	312.8
Noord-Brabant	203.4	609.4	406.0

Source: de Cooman *et al.*, 1995

Table 3.3 Total number of holdings represented by FADN and holdings with supply levels of animal manure exceeding 170 kg of nitrogen per hectare by member states in 1990/91

Country	Total number of holdings (× 1000)	Average supply of animal manure (kg N/ha)	Holdings with nitrogen from manure > 170 kg/ha		
			Share of total number of holdings (%)	Share of total manure production (%)	Average supply (kg N/ha)
Belgium	51.9	196	47	71	327
Denmark	81.0	109	26	59	258
Germany	373.9	98	12	21	207
Greece	498.3	64	15	68	557
Spain	690.6	40	19	64	723
France	556.7	62	6	18	309
Ireland	140.2	93	8	17	225
Italy	369.8	55	6	49	361
Luxembourg	2.3	128	11	15	197
Netherlands	94.0	343	63	99	501
Portugal	448.5	40	18	35	357
United Kingdom	141.6	68	17	27	258
EUR 12	4448.9	73	13	40	352

Source: Brouwer *et al.*, 1996

170 kg N/ha shows that 100% of the grazing livestock farms and 100% of the granivore farms exceed this norm, as do 84% of the mixed farms (Table 3.4).

In Table 3.5 the quantities of mineral fertilizer and organic manure have been added and are compared to equilibrium fertilization units. The results show that all of the provinces exceed the limit: for Zeeland and Flevoland this is small, but Noord-Brabant shows a surplus of more than 500 kg N/ha/year.

Other sources of N

Other sources of nitrogen input to the soil in the Netherlands are sewage sludge, compost, atmospheric deposition and fixation by crops. Agriculture is by far the biggest contributor of N to the soil and thus to surface and groundwater.

Nitrogen balance

Since both the level of utilization of mineral fertilizer and the production of animal manure are very high in the Netherlands, it will come as no surprise that the mineral balance of input (manure and fertilizer plus depositions from the atmosphere) and uptake is 321 kg N/ha for the country as a whole, a higher surplus than any other EU country. Within the Netherlands, variations occur between the different provinces, ranging from 489 kg N/ha in Noord Brabant to 100 for Flevoland (see Table 1.3). Transport of manure to other parts of the country has

Table 3.4 Farms with manure production exceeding 170 kg N/ha and share in total number of holdings (%) by farming type in 1990/91[a]

Country	Grazing livestock farms		Arable farms		Mixed farms	
	Total (×1 000)	Share (%)	Total (×1 000)	Share (%)	Total (×1 000)	Share (%)
Belgium	12.9	68	3.6	100	7.7	54
Denmark	10.7	68	4.1	88	5.8	28
Germany	23.2	15	3.6	84	16.5	14
Greece	42.5	83	0.9	100	12.2	32
Spain	88.1	54	13.4	84	25.8	32
France	11.7	5	8.5	97	11.1	12
Ireland	10.2	8	.	.	.	.
Italy	50.9	33	4.2	86	18.6	13
Luxembourg	0.2	11	.	.	.	.
Netherlands	42.6	100	9.7	100	6.9	84
Portugal	20.9	30	3.0	95	37.2	26
United Kingdom	16.7	21	4.3	100	1.9	13
EUR 12	330.5	30	55.7	92	144.0	21

[a] When the minimum threshold of 15 farms for the sample size if not reached for a farming type, no data are given.

Source: Brouwer et al., 1996

not been taken into account in these figures, which means that in actuality, surpluses per province will be more evenly spread than the above numbers show. Figure 3.1 presents quite a dramatic picture of regional nitrogen balances and the position

Table 3.5 Organic and mineral fertilizer (kg N/ha per year)

NUTS II	Limit	Utilization	Difference
Zeeland	210.0	253.2	43.2
Flevoland	203.0	269.6	66.6
Groningen	265.1	373.8	108.6
Noord-Holland	322.1	430.9	108.8
Drenthe	285.2	399.0	113.9
Zuid-Holland	317.8	433.1	115.4
Friesland	387.7	611.4	223.7
Utrecht	385.6	693.1	307.5
Overijssel	350.5	700.2	349.7
Gelderland	339.8	748.0	408.2
Limburg	258.0	682.9	424.9
Noord-Brabant	282.1	808.3	526.2

Source: de Cooman et al., 1995

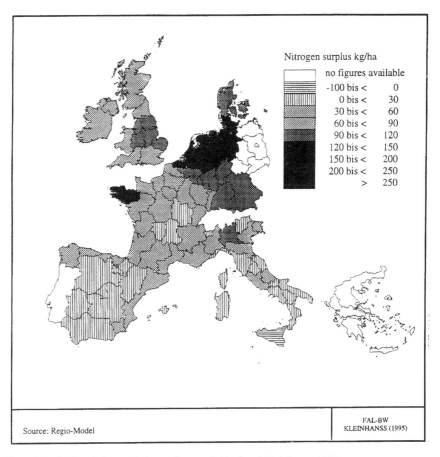

Figure 3.1   Regional nitrogen balance. Source: Schleef and Kleinhauss, 1996

of the Netherlands: together with the Flemish part of Belgium, it has the highest nitrogen surpluses in the European Union.

Phosphorus
The nominal annual phosphate production in agriculture, as calculated by the Ministry of Agriculture, is shown in Table 3.6. In 1987, control measures were started. The nominal and estimated production figures show a steady decline since that year. Agriculture is responsible for 90% of the total annual phosphorus input into the Dutch soil; its contribution to surface water was relatively small (18%) in 1985, but has become relatively larger, since contributions from other sources have decreased more rapidly (Table 3.7). In absolute terms, there has been little change in that period.

Table 3.6   Phosphates from animal manure

Year	Phosphate production (thousand tonnes)
1970	176
1980	244
1990	273
1996[a]	206
1998[a]	200

Source: Eurostat

[a] Estimates

Table 3.7   Sources of phosphate input into surface waters in the Netherlands, 1985–1993

Source	Contribution, totals per year and % by source		
	(33 kton P) 1985	(26 kton P) 1990	(17 kton P) 1990/1993
Communal	39	27	36
Industry	42	46	23
Diffuse/Agriculture	18	27	40 (1990)

Source: RWS/RIZA

AMBIENT ENVIRONMENTAL CONDITIONS OF SOILS, SURFACE-, GROUND- AND DRINKING WATER

Soil conditions

In the Netherlands, phosphate saturation of soils is a major problem particularly in the sandy soils in the eastern and southern parts of the country. This means that, even if phosphate input is reduced in the coming years, phosphates will leach from these soils for a long time to come: present reductions in input may not have an effect on phosphate leaching from these soils for 30 or 40 years. There is, therefore, no direct relationship between phosphate input and leaching.

Soil conditions are also important for the leaching of nitrates. High groundwater levels promote denitrification; on the other hand, low ground water levels are related to lower levels of nitrate leaching into surface waters. As a result of manure transports from areas with high manure production levels (sandy soil areas in the east and south) to areas with lower levels in the western part of the country have resulted in lower nitrate leaching in the east and south, but increased levels in the western part.

Surface water

Reports on the quality of surface water in the Netherlands are published yearly. Water quality is measured at several sampling points, both in running and in stagnant waters, and then compared to norms and parameters for eutrophication: 2.2 mg N-total/l. Table 3.8 shows some results from points where a minimum of six samples have been taken during the summer. Table 3.9 shows that the relative contribution of N into surface waters has increased somewhat between 1985 and 1990/93, while the total input has decreased about 5%, all due to the lower input from industrial sources.

As Figure 3.2 shows, there is a big difference in the level of nitrate pollution of surface waters as a result of run-off in sandy regions and those in areas with clay or peat soils. In the sandy regions in the east and southeast (Twente, Oost Gelderland and Oost Brabant) the nitrate concentrations are very high, while in the regions with clay or peat soils in the north, northeast and western parts of the country these concentrations are only moderate (Fig. 3.2).

Virtually all surface water in the Netherlands is at risk from phosphate pollution (Fig. 3.3). The load is determined by point discharges, such as sewage treatment works, industry and non-point agricultural sources. As Table 3.7 shows, the relative contribution of agriculture to phosphate pollution in surface water in the Netherlands has increased since 1985, because phosphates from communal sources and industry have decreased more rapidly. The total amount of phosphate input into surface waters had decreased considerably, but the decrease from agricultural sources has been relatively minor.

Where soils are saturated with phosphate, leaching from shallow groundwater can play an important role. It is not unusual for watercourses in areas with sandy soils to have phosphate concentrations in the region of 2 mg P/l. About 80% of the small streams in the country exceed the environmental quality objective for phosphate of 0.15 mg P/l.

Lakes and marine waters

The N concentration in the Ijsselmeer was 4.39 mg/l in 1980, 4.14 in 1985 and 3.95 in 1990. As far as the North Sea is concerned, studies undertaken for the Paris Commission show that phosphorus is the limiting factor for the development of algae in the coastal waters of the Netherlands, up to 30 km from the coast: further out nitrogen becomes the limiting nutrient (see Figure 3.4).

Table 3.8 Water quality of surface waters, compared with the norm of 2.2 mg N/l

	Number of sampling points	Exceeding the norm in 1992	Exceeding the norm in 1991
All surface waters	146	89%	77%
Stagnant waters	45	84%	74%

Table 3.9 Input of N into surface waters in the Netherlands from different sources

Source	1985 100% = 220 kton N	1990 100% = 210 kton N	1990/93 100% = 200 kton N
Commual	21%	21%	21%
Industry	7%	6%	2%
Diffuse/Agriculture	72%	73%	76% (1990)

Source: RWS/RIZA

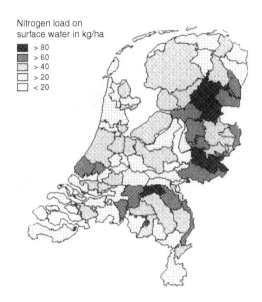

Figure 3.2 Nitrogen load on surface water due to leaching and run-off in 1993 by hydrological district. Source: RIZA

Groundwater

The average concentrations of nitrate (measured as nitrogen) found in groundwater and differentiated by use and soil type, are shown in Table 3.10. The considerable differences between nitrate level applied to the soil and the concentrations that are found in the groundwater in regions with clay and silty soils, can be explained by the important process of denitrification. In fact, the slow infiltration of water into the subsoil and a groundwater level that is not very deep, create conditions that are favourable for denitrification.

Figure 3.5 gives an even better picture of regional differences in the nitrate inputs to groundwater. This figure indicates the estimated nitrogen concentrations of the water percolating into the subsoil. In the Netherlands, pollution of groundwater by nitrate is a very serious problem, because roughly two-thirds of

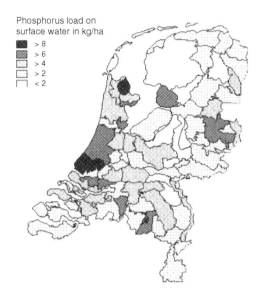

Phosphorus load on
surface water in kg/ha

- ▓ > 8
- ▓ > 6
- ☐ > 4
- ☐ > 2
- ☐ < 2

Figure 3.3 Phosphorus load on surface water due to leaching and run-off in 1993 by hydrological district. Source: RIZA

the drinking water supplies are abstracted from groundwater. The other third is supplied, directly or indirectly, by the river Rhine and the Meuse. So far, groundwater reserves have provided a high quality water supply that requires only a minimal treatment to make it suitable for human consumption. Now, increasing nitrate concentrations of aquifers are creating problems: at a small number of extraction points the nitrate concentration in groundwater has exceeded the value of 50 mg/l, the EU limit for drinking water.

Nutrient losses due to agricultural activities
The contribution of agriculture to nitrate pollution of surface waters has been estimated at 76% of the total pollution due to human activity. In absolute numbers, this represents 152 000 tons N/year of agricultural origin, versus 200 000 tons N/yr total (Table 3.9). There are no reliable estimates for ground water.

NEED AND PURPOSE OF EXISTING POLICY AND REGULATION (INTERNATIONAL AND NATIONAL)

International policy
A number of international policy decisions regarding water quality of ground- and surface waters have been agreed upon, in which the Netherlands has been a participant, like INSC and the Paris Convention. The Netherlands has accepted all decisions, rules, measures and initiatives agreed upon by the International

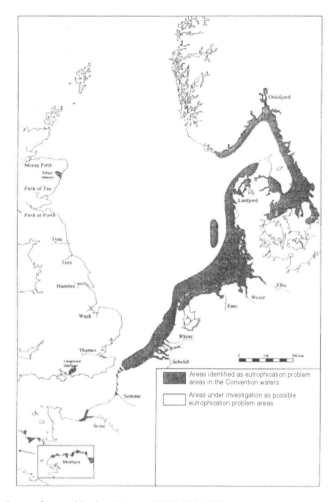

Figure 3.4 Zones of eutrophication. Source: PARCOM, 1990

Northsea Minister Conferences (1984, 1985, 1990). It has also accepted all PARCOM recommendations (1988, 1989, 1992, 1995) to reduce emissions of nutrients to the North Sea by 50%.

The Nitrate Directive of the European Union has not been transposed (6/96), but

- Vulnerable zones have been designated (whole territory designated),
- A Code of good agricultural practice has been notified, and
- Action programs have been notified, but have been withdrawn temporarily in November 1996.

Table 3.10 Average concentrations of nitrate in groundwater, 1990 (1992)

Soil Type	Use	Concentration (mg N/l) Shallow groundwater	Concentration (mg N/l) 10 m.	
Clay	Arable	22	2.8	
	Grassland	3	0.5	
	Natural	3	2.5	
Peat	Grassland	0.6	0.3	
	Natural	0.8	0.3	
Sandy	Arable	63	23 ('90)	25 ('92)
	Grassland	24	2.6	
	Natural	2	2.2	2.9

Source: RIVM

National policy

The need for a national policy on minerals will be apparent from the data on pressures from minerals (especially phosphate), sources and ambient conditions that have been presented above. The very intensive nature of agriculture, especially livestock farming, high fertilizer use and manure surpluses, puts a heavy environmental burden on surface- and groundwater and contributes to soil saturation. The other side of the coin is the importance of agriculture to the Dutch economy, which needs to be balanced against environmental concerns. Thus, the overall policy goal of the Dutch government has been to maintain an economically viable agriculture, while at the same time reducing the heavy burden on the environment. Protection of the environment has focused on the prevention of eutrophication (reduction of nitrates and phosphates), soil saturation (phosphate) and groundwater protection (nitrate reduction), prevention of acidification (ammonia emissions) and of negative effects of nutrients on nature.

In the mid-1980s, the need for more environmental regulation became apparent, as the dangers of eutrophication and groundwater pollution became national and international concerns (INSC, PARCOM). In the context of these international agreements to reduce the emission of nutrients, the overall medium-term goal of the policies of the Dutch government concerning mineral surpluses, have been to reduce the 1985 emissions of phosphate and nitrate from all sources to surface waters by 50% and to limit the leaching of nitrate from the topsoil to groundwater to the extent that groundwater abstracted for drinking water supplies complies with the EU limit value of 50 mg NO_3/l.

The EU Nitrate Directive formulated specific manure and fertilizer policy goals: not more than 170 kg N/ha per year can be applied to the soil in the year 2000. These goals must be implemented by the following means: designation of vulnerable areas, formulation of a Code of Good Agricultural Practice and an Action Program, that specifies by what means these goals will be met (see also Chapter 2).

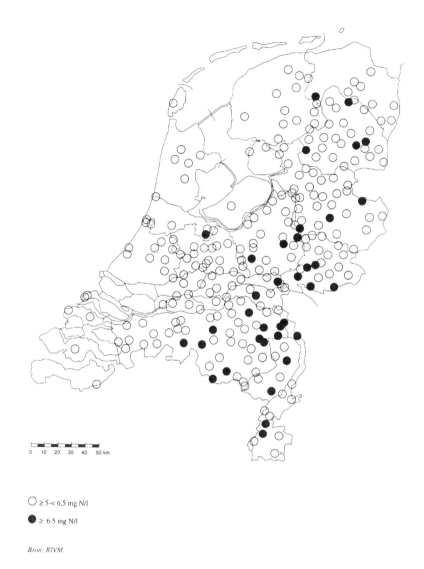

Figure 3.5 Concentration of N in superficial groundwater. Source: RIVM

Because the Dutch government realized that drastic changes would have to be made, but that this would not be possible overnight, a decision was made to implement the new policies in three stages. A gradual change process was envisioned, to give farmers a chance to adjust to new environmental demands, that would be different for each phase. Each of the three phases had a distinct objective:

Phase 1 (1987–1990): Stabilization

The goal of the initial phase was the stabilization of manure production and utilization at a level where all manure produced could be applied nationally, to prevent a nationwide manure surplus. The decision was made that in phases 1 and 2 the production and application of manure would be controlled by phosphate regulation, in contrast to the other EU countries. This decision caused some controversy, since nitrate causes more serious problems in agricultural soils. However the decision was based on the fact that phosphate is easier to quantify. Nitrate content in manure varies with circumstances such as age, method of application and weather, and soil conditions can greatly influence the amount of nitrate that will ultimately leach into the groundwater. In regulating N through P, a fixed ratio in manure between the two is assumed, presently 2.6:1.

This phase was primarily directed towards stabilizing the problem of mineral surpluses by adjusting the composition of animal feeds, manure distribution and manure processing. Manure production rights (Manure and Fertilizer Act) and use standards for livestock manure were introduced (Soil Protection Act). These standards were set at such a level that manure surpluses at a national level could be prevented, while 'manure banks' would redistribute excess animal wastes.

Phase 2 (1990–1994): Tightening standards

Policy objectives for the second phase were to avoid further accumulation of nitrate in soil and water, by tightening use standards for the application of fertilizers in agriculture gradually, until an equilibrium between input and output has been achieved, a policy goal that was to be achieved in phase 3.

This phase was geared towards an actual decrease of the pressures on the environment. Use standards were tightened and reduction targets were set in such a way that farmers would have sufficient time to work towards meeting the new standards. It was important to avoid the development of a nationwide manure surplus, that cannot be utilized at all. In order to meet these new standards, the development of environmentally friendly technology and new management practices was stimulated.

Phase 3 (1995–2000): Equilibrium

The policy goal is that by the end of the third phase, the final objective of Dutch manure and ammonia policy, the equilibrium between inputs and outputs, will be achieved. Central to this policy, outlined in the Policy Memorandum on Manure and Ammonia (Phase 3), is the reversal from a generic to a farm-oriented approach. Instead of the application standards, which make no allowances for difference between farms, 'standards for acceptable losses' were proposed. These loss standards, which apply to both P and N, establish levels of total allowable surpluses, that are the result of inputs of organic and mineral fertiliser minus outputs in crops and other agricultural products. In this manner, the envisaged equilibrium between inputs and outputs would have to be realized. In order to realize these loss standards at farm level, a new accounting system (MINAS) would have to be implemented. Levies would have to be paid on surpluses exceeding the standard. The government and the industry agreed that the feasibility of the proposed loss standards should be investigated by a study.

When the Policy Memorandum was worked out into concrete measures, problems became apparent: the change-over to the new accounting system would take more time, manure reprocessing developments were behind schedule and the study showed that the gap between environmentally acceptable losses and losses that occur with good agricultural practice, could not be bridged. This meant that the policy goal of an equilibrium between input and output in 2000 was unrealistic.

A new policy document (Policy Document on Manure and Ammonia) was presented to the Second Chamber in October 1995. In February 1997, several amendments were made, after which it was accepted. The main points are:

- The Minerals Accounting System (MINAS) will enter into force in 1998 (not '96).
- Loss standards will be tightened gradually between 1998 and 2008/2010.
- Initially, only farms with high livestock densities are obliged to use the Minerals Accounting System; later this requirement will apply to all farms.
- For the first 2 years, levies will be low and will only apply to phosphate surpluses from manure, not from mineral fertilizer; the acceptable loss norm for nitrogen may be reduced by a correction factor for ammonia emissions that take place. In practice this will mean that levies will be very low between 1998 and 2000.
- In order to reduce manure surpluses, the government will siphon off manure production rights; however, there has been a regional initiative to achieve an extra reduction of the minerals in manure through lower mineral content in feedstuffs, instead of the government siphoning of manure production rights. This initiative may be tried and then evaluated after 2 years.
- The policy on ammonia aims to reduce ammonia emissions 70% by 2005, compared with 1980. This objective will be realized by low emission application of manure, low-emission housing and additional regional ammonia policies, f.e. in nature areas.

In the last paragraph of this chapter, these proposals will be presented in more detail.

POLICY FORMULATION AND LEGAL INCORPORATION

The first phase of the Dutch policy relating to agricultural pollution, is incorporated in the following legislation: the Manure and Fertilizer Act, the Soil Protection Act, the Surface Water Pollution Act, the Nuisance Act and Provincial regulations.

The Manure and Fertilizer Act
The Manure and Fertilizer Act (1986) has four main objectives:

1. To control the composition of fertilizers.
2. To introduce a licensing system for trade in certain types of fertilizer derived from waste products, such as sludge and compost.

3. To impose limits on the maximum amounts of certain fertilizers (use standards).
4. To regulate the disposal of excess manure.

Two features of the Act deserve special mention. The Act provided for the establishment of 'manure banks', facilities intended as regional depots for storage, processing, destruction or redistribution of excess animal wastes. Manure banks can redistribute local manure surpluses to other parts of the country, where it can be applied on for example arable farms. Second, it introduces a system of charges on surplus manure. These charges are to be levied on the amount of phosphate produced by livestock. Farmers are required to maintain a register of the amount of manure produced on the farm. The Act also prohibits the establishment of new holdings or the expansion of existing farms above the point where the production of phosphate through manure exceeds 125 kg/ha/year (manure production rights).

The Soil Protection Act
The objective of the Soil Protection Act (1986) is to introduce a national soil protection regime for all activities which threaten to pollute or damage the soil or groundwater. The Use of Animal Manure Decree controls the application of manure. It regulates the maximum application allowance of manure, depending on the crop grown, the period of application and the periods during which and conditions under which fertilizers may be applied. It also designates areas in the sandy soil regions of the country, that are most vulnerable for nitrate leaching, where the period of application of manure is shorter than in other parts of the country.

The Soil Protection Act makes two further provisions, which are of special interest. First, all 12 provinces are required to designate soil protection areas, where special protection for the abstraction of drinking water supplies is called for. These designations and the protective measures are to be laid down in provincial ordinances. Second, provinces should also designate soil protection areas, where the physical, chemical and biological characteristics of the soil have only been influenced to a minimal degree by human activities. These areas are to be designated in programs especially drawn up for the purpose.

The Surface Waters Pollution Act
The Surface Waters Pollution Act of 1969 controls polluting substances that are so disposed of that it can be foreseen that they will pollute surface waters. It controls direct discharges, pollution of land (expected to be) inundated and run-off. It implicitly prohibits application of fertilizer close to water courses.

The Environmental Management Act (1993)
The Environmental Management Act (1993) is the successor to the Nuisance Act of 1983. It provides that activities that are likely to cause danger, damage or nuisance to others and to the environment are to be licensed, usually by the

municipality, but in special cases by the Provinces. It sets standards for establishing new or renewed farms. It also specifies requirements for installations and buildings. Virtually all livestock farms require an environmental licence from the municipality. Two issues are important for obtaining a licence: the first one is odour, the second is ammonia deposition in nature areas. Livestock farms that are close to ecologically valuable areas are required to make special adjustments. There is also a decree for manure stores, which specifies demands for its construction and how it should be covered, also to reduce ammonia emissions. Municipalities located in areas with many intensive livestock farms (so called 'concentration areas' in the provinces of Overijssel, Gelderland, Utrecht, Noord-Brabant and Limburg) develop ammonia reduction plans, required by the Interim Act Ammonia and Livestock Farming.

The Manure Production Relocation Act (1994)

The Manure and Fertilizer Act (1986) contains a Relocation Decree (1987), which states that it is possible under clearly defined conditions, to transfer the right to phosphate production to another location or to retain its 'reference quantity' of phosphate (its registered 1986 production), when relocating to another area. Under the Manure Production Relocation Act (1994), 25% of manure production rights will be siphoned off when these rights are sold or moved to another location. When manure production rights are transferred within the family and the farm is continued in the same location, no siphoning off will take place.

REGULATORY CONTROLS

Regulatory controls on agricultural practices relating to livestock waste, fall into three categories: control measures relating to the production of manure, restrictions on the application of manure and fertilizer to the land and the disposal of surplus livestock wastes.

Measures relating to production of manure

In the first category, there are three different measures:

1. The imposition of a 'surplus levy' on the quantity of phosphate (P_2O_5) produced by livestock, which exceeds 125 kg phosphate/ha each year. Under the Manure and Fertilizers Act, the first 125 kg/ha each year are free, quantities over that amount are charged. The calculation of the phosphate produced by livestock is regulated by the Designation of Animal Species and their Manure Production Decree. This lists various breeds of cattle, pigs and poultry and lays down a nominal production for each animal.
2. A second way of limiting the nutrient surplus is through adjustments in the composition of animal feed. Especially for pigs and poultry, this is a possibility to reduce the mineral content of manure. For these sectors, the Mineral Supply Registration System has been introduced. When low-mineral foodstuffs are

used, lower nominal values can be used. If the farmer can show that his manure contains less phosphate, he will pay lower levies on the surplus.

3. A third way of limiting the quantity of manure is to impose restrictions on the phosphate production of farms. Phosphate production will be limited to 125 kg/ha each year on new farms. Existing farms cannot expand beyond this limit. Existing holdings with a phosphate production above this limit on 31 December 1986 are prohibited from expanding further, but are entitled to maintain manure production at this level: manure production rights. These rights can be transferred under clearly defined conditions: if they are transferred to another area or are merged with another holding, 25% of the manure production rights will be siphoned off; if the rights are transferred within the family, the holding maintains its 'reference quantity'.

Buying back manure production rights from farmers is a way of decreasing total manure production and thus of preventing a nationwide manure surplus from occurring. However, the net result is also a curtailment of the livestock sector and thus a loss of employment and productivity. A proposal of the government to increase the rate of siphoning off manure production rights, was unacceptable to the agricultural sector and has since been dropped.

Measures that control the application of nutrients

Limits on the quantity of phosphate
In the first place, limits have been set on the maximum quantity of phosphate in manure and other organic fertilizers, which may be applied to agricultural land. These limits depend on the type of crop grown; they will be progressively tightened, until an equilibrium between input and output has been reached. Two aspects of these limits deserve special mention:

● There has been some controversy in the Netherlands about the decision to base these limits on phosphate rather than nitrate, since the latter causes more serious problems in agricultural soils. However, an important consideration has been the fact that phosphate is easier to quantify. Nitrate is harder to quantify because of ammonia emissions and denitrification. In the new Mineral Accounting System (see below), both P and N have to be accounted for.
● Until now, the Netherlands have utilized 'application standards': maximum permissible quantities of phosphate in manure, which may be applied to agricultural land. The application standard for phosphate in manure and other organic fertilisers for grassland for 1996 is lowered from 150 to 135 kg/ha per year.

The new Minerals Accounting System will become obligatory in 1998 for farms with a livestock density of 2.5 LU/ha. It is based on 'acceptable losses' of both N and P and applies to manure and inorganic fertilizer, even though phosphate in fertilizer will not be subject to levies until 2000. In 2000 this system will be required for all livestock farms and eventually it will presumably apply for all agricultural farms (Table 3.11).

If no manure is produced at the farm, the application standards presently utilized will be replaced by phosphate supply standards in 1998. In 1998 a maximum of 120 kg/ha of phosphate may be supplied in the form of manure and other organic fertilizer on grassland, 100 kg/ha on arable land. These supply standards will be lowered to 80 kg/ha in 2002 (Table 3.11).

Restrictions on when and where to apply manure
Other control measures impose restrictions on the period during which manure can be applied to the land. The intention is to limit the input of P and N outside the growing season when crops cannot take up nutrients. These periods are:

Arable land on sandy soils*   01/09–10/02
Grassland on sandy soils*   01/09–01/02
Other grasslands   15/09–01/02
* As designated in the Use of Animal Manure Decree

Restrictions also apply to the application of manure on soil covered with snow, near water courses (use of special equipment recommended) and in designated ground water protection areas. Manure application in nature reserves is prohibited. Other restrictions apply to the application of manure in order to minimize ammonia emissions by means by manure injection or ploughing within 12 hours of application.

Management agreements
Special management agreements can be used in the interest of the conservation of nature and landscape, between farmers and the provinces. A total area of

Table 3.11   Acceptable loss standards in kg/ha and phosphate supply norms in kg/ha

Minerals, kg/ha	1998	2000	2002	2005	2008/2010
MINAS farms *					
Phosphate	40	35	30	25	20
Nitrogen grassland	300	275	250	200	180
Nitrogen arable land	175	150	125	110	100
Non-MINAS farms **					
P-supply/ha					
Grassland	120	85	80	80	80
Arable land	100	85	80	80	80

* Farms under obligation to utilize MINAS; in 1998 the norm is over 2.5 LU/ha, but in 2000 this limit will be lowered to LU/ha.
** Supply norms for phosphate apply to farms below 2.5 LU/ha (2 LU in 2000)

200 000 ha has been designated. The two main instruments in these areas are management agreements and the establishment of nature reserves. Farmers are compensated for their efforts.

Measures relating to the disposal of livestock wastes

Storage facilities
After the application of manure was prohibited during the fall and winter, it became necessary to increase the existing storage capacity for livestock wastes. In the Netherlands, the total capacity of the storage facilities is not regulated by law, but is a natural consequence of the periods when manure application is prohibited. The total capacity should be up to 9 months' supply. Subsidies for the construction of new and improved storage facilities are available.

Manure banks
The primary function of the manure banks is to act as an intermediary in routing livestock wastes from farms with a surplus to farms which wish to use the manure as fertilizer. In performing this task, the manure bank only puts suppliers in contact with traders in livestock waste; it does not itself handle the surplus. This only occurs when it acts as a depot for surplus livestock waste, for which farmers have been unable to find an outlet. Farmers who wish to use the manure bank in this capacity, are required to pay a fixed rate for disposal, with the result that the bank is not used much for this purpose. The role of the manure banks will be reduced after 1/1/98. If the redistribution strategy is to be successful, then the quality of the wastes offered as fertilizer is of great importance. The prices of the premiums that have to be paid for different quality manure are lower for higher percentages of dry matter. Lately, farmers have invested in facilities for drying manure, to improve quality and lower premiums.

Reprocessing plants
The development of reprocessing plants was an ambitious plan of the Dutch government to solve the manure surplus problem. However, it has now become clear that large scale processing of manure is too expensive to make it a feasible option. Plans of the government to subsidize the operation of the plants have been rejected by the European Commission as unacceptable government aid, so presently, large scale manure processing is no longer seen as a solution to the mineral surplus in the Netherlands.

All these measures have been included in the Code of Good Agricultural Practice, required by the EU in the context of the Nitrate Directive.

MONITORING, CONTROL, RECORD KEEPING AND VERIFICATION

Groundwater and surface water monitoring
Networks for monitoring groundwater and surface water quality have been in operation for a long time in the Netherlands. The national groundwater monitoring

network consists of about 490 measuring stations. Measurements are taken at depths of 10 m, 15 m and 25 m and the parameters include both phosphate and nitrate. There are also around 400 provincial measuring stations, which partly overlap the national network.

Monitoring and control of production, application and disposal of manure
Three organizations exercise control over the production, application and disposal of livestock wastes: the General Inspection Service, the Levies Bureau (deals specifically with manure registers and surplus levy) and the police. Most important is the General Inspection Service of the Ministry of Agriculture (AID). One of its five divisions is specifically concerned with environmental protection, enforcing the Manure and Fertilizers Act, the Soil Protection Act, the Pesticides Act and legislation relating to nature protection and hunting.

In 1987, the Ministry of Agriculture required statements from about 90 000 livestock farmers about their livestock in order to fix the quantity of phosphate produced on each farm in 1986. This 'reference quality' fixed the maximum phosphate production allowable for each farm, rendering it important for farmers to secure as high a reference quality as possible. The AID investigated 7000 farms and found serious discrepancies in 1800 cases (38%).

The Manure Registration System (MAR) is presently the most important instrument for controlling the production, application and disposal of manure. Livestock farmers with a phosphate reference quantity about 110 kg/ha are required to maintain a manure register, as are all livestock farmers in groundwater protection areas. The manure registration system has three objectives:

● to record manure production, application and disposal on each farm;
● to provide information on establishment, expansion and relocation;
● to provide a basis for the calculation of the surplus levy.

The system met a lot of resistance from farmers from the beginning. In the first year, about one-third of the qualifying farmers refused to cooperate, in 1988 this dropped to one-fifth. Random inspections found one in every ten farmers to be in breach of legislation. As a result, registers have been simplified: an initial statement of the number of animals kept, the area of the farm and the reference quantity of manure is required. Only subsequent changes need to be notified to the Levies Bureau. In 1998 the manure registration system will be replaced by the Minerals Accounting System.

Transfers of livestock wastes also need to be registered in a delivery certificate. This certificate records the quantity of manure transferred, the name of the supplier, the transporter and the buyer. Inspectors of the manure banks carry out random checks, the AID is involved in cases of suspected fraud. Nevertheless, an evaluation of the manure policies found that in 1987 32 000 tonnes of manure were unaccounted for; in 1988 that amount had risen to 446 000 tonnes.

Finally, there are also indications that control by municipalities of Environmental licenses, is not as strict as it should be. Many livestock farmers may have increased the number of animals they keep without applying for a new licence, as the Environmental Management Act requires.

FINANCIAL MEASURES AND INCENTIVES

Financial measures and incentives play an important role in the implementation of different policies. Levies are charged for:

- Surplus manure production: the quantity of phosphate produced by livestock, exceeding the standard of 125 kg/ha per year. The levy was Nfl. 0.25/kg for the next 75 kg/ha, any quantities over 200 kg/ha were charged at a rate of Nfl. 0.50/kg. After 1998, this levy will be replaced by a levy on exceeding allowable losses, initially only on phosphate loss, eventually also on nitrogen losses (Table 3.12).
- Disposal of waste to manure bank: fixed rate charge of Nfl. $20/m^3$.
- Destination levy in the southern and eastern part of the country.

Financial support is available for:

- Innovative techniques to reduce the production of manure or improve reprocessing or disposal of livestock wastes (Bijdrageregeling praktijkprojecten mestproblematiek): 20–40% of the total investment.
- Investments of contractors which carry out work on the farm for purposes of environmental improvement; subsidies of 20–25% (until 1994).
- Investment in expanding storage capacity for livestock wastes and for sealing them.
- Investments in joint manure storage facilities.
- Investments in manure reprocessing plants.

There has been an initiative to reward those farmers who stay below the acceptable loss norms with a premium, but this issue remains to be decided.

Farmers in groundwater protection areas are entitled to some form of compensation for the restrictions imposed upon them. These compensations will gradually disappear, when equilibrium fertilization has been achieved, because controls in groundwater protection areas will be the same as outside the area.

Compensation can also be granted for management agreements, whereby the

Table 3.12 Levies on mineral surpluses

Mineral Surplus	1998–2000	After 2000*
Phosphate, 0–10 kg/ha	Dfl. 2,50	Dfl. 5, -
Phosphate, >10 kg/ha	Dfl. 10, -	Dfl. 20, -
Nitrogen, per kg/ha	Dfl. 1,50	Dfl. 1,50

* Levy amounts for 2000 are tentative and may be altered after evaluation

farmers agree to implement certain measures in the interest of nature conservation. Nine classes of management agreements are recognized, with their own level of compensation.

EFFECT OF MEASURES

Little is known about the environmental effects of all these efforts on the part of the Dutch government to reduce emissions of phosphate and nitrate to surface water and reduce leaching of nitrate to groundwater. PARCOM has published some comparative data on the application rates of manure and mineral fertilizer and the N and P surplus of different countries between 1985 and 1992 and the results show that:

- In the Netherlands, application of total-N in manure decreased from 296 to 209 kg/ha/year, for total P this was from 53 to 44 kg/ha/year (Table 3.13).
- Application of total-N in mineral fertilizer decreased 23%, P-total 11% (Table 3.14).
- The total-N surplus was reduced 19% between 1985 and 1990, the total-P surplus 19% between 1985 and 1992 (Table 3.15).
- Total inputs of total-N between 1985 and 1992 show a reduction of 2%; provisional data based on modeling studies show a reduction of 15% total-N between 1985 and 1995. For P, measured as phosphate, the rates are -5% between 1995 and 1992, 0% between 1985 and 1995 (Table 3.16).

Table 3.13 Average application rates of manure (kg/ha/year) 1985–1992.

Country	Total-N 1985	Total-P 1985	Total-N 1992	Total-P 1992
Belgium	222	50	252	53
Denmark	100	20	91	17
France	46	10	46 (1990)	10 (1990)
Germany	73.5	22.3	86.3	16.1
Original Federal States	78	23.9		
New Federal States	18.5			
Iceland[a]	21	7	20	7
Netherlands	296	53	209	44
Norway	48	8	44	8
Sweden	35[b]	7[c]	25 (1993)	8 (1993)
Switzerland[d]	149	24[d]	126 (1990)	21 (1990)
United Kingdom	74	17	72	18

[a] These figures are based in an estimated of the manure collected and applied in Iceland and and estimated 130 000 ha of total cultivated land.

[b] Total-N 1986

[c] Total-P 1986

[d] Per hectare of land used (crops and grassland) without alpine pastures, para-agriculture enclosed.

Source: OSPAR 1995

Table 3.14 Application rates of chemical nitrogen and phosphorus fertilisers 1985–1992.

Country	Total-N		% Reduction 1985–1992	Total-P		% Reduction 1985–1992
	Fertiliser used (kg/ha/year)			Fertiliser used kg/ha/year		
	1985	1992		1985	1992	
Belgium	159	146	8	27	20.5	24
Denmark[a]	140	132	6	17	12	29
France	81	92	−13.5	24.2	19.6	19
Germany[b]	125	94 (1993)	25% (1985–1993)	25.2	10.6 (1993)[c]	58% (1985–1993)
Iceland[d]	104	88	15	25	18	28
Netherlands	250	192	23	18	16	11
Norway	121	118	2	22	18	18
Sweden	84	75	11	16	8	50
Great Britain	144	124 (1993)	14 (1985–1993)	18	15 (1993)	17 (1985–1993)
Switzerland[e]	59	59 (1990)	0 (1985–1990)	15	14 (1990)	7 (1985–1990)

[a] Per hectare of land used.

[b] Quantites reported relate to the Original and the New federal States.

[c] Preliminary results for 1992/1993.

[d] Figures are based on the total amount of chemical nitrogen and phosphorus fertilisers sold in Iceland and the estimate of total cultivated land area of 130 000 hectares.

[e] Per hectare of land used (crops and grassland) without alpine pastures, para-agriculture included.

Source: OSPAR 1995

- More recent data show that in 1994 and 1995, nitrogen surpluses increased: the N-surplus increase in 1994 was 2%, a provisional estimate for 1995 is 5%. These increases are partly due to weather conditions (high temperatures and little rain, resulting in a lower uptake of nutrients by crops), partly to a higher utilization of mineral fertilizer (9%) and animal manure (2%). The phosphate surplus increased 3% in 1994, also due to the weather, but decreased 10% in 1995 (provisional data), due to a lower utilization of mineral fertilizer (10%) and of animal manure (5%). The data for N reverse a trend of a gradually decreasing surplus since 1985 (Fig. 3.6; Fong, 1996). Another study (Van Eerdt, 1996) shows that phosphate excretion has decreased 5% for all livestock, but relatively more for pigs (26% since 1986) and poultry (14%), while nitrogen excretion has remained at the same level.
- The data in Table 3.8 shows that the contributions of P and N from agriculture to surface waters have decreased about 5% for P, with little change for N.

The differences in these results are difficult to interpret, because they are not directly compatible; however, they do suggest moderate levels of reduction over a 7–10 year period, with a slight increase in N-surplus in recent years. The reductions that have

Table 3.15 Nitrogen and phosphorus surplus 1985–1992.

Country	Total-N			Total-P		
	Surplus kg/ha		% Reduction 1985–1992	Surplus kg/ha		% Reduction 1985–1992
	1985	1992		1985	1992	
Belgium	195	207	–6	37	35	5.5
Denmark	160	160 (1990)	0 (1985–1990)	20	20 (1990)	0 (1985–1990)
France	50–60	50–60 (1990)	0 (1985–1990)	15–20	15–20 (1990)	0 (1985–1990)
Germany	111[a]	103 (2)	7%	20 (1)	6,5 (2)	67
Netherlands	409	332 (1990)	19 (1985–1990)	48	39	19
Norway[c]	105	87 (4)	17	19	11 (4)	42
Sweden	90	78 (1990)	13 (1985–1990)	15	12 (1990)	20 (1985–1990)
Switzerland[a]	146	119 (1990)	18	20	16 (1990)	20
United Kingdom	142[e]	126 (1993)[e]	11 (1985–1993)	14.2[e]	12.4 (1993)[e]	15 (1985–1993)
	92[c]	77 (1993)[f]	16 (1985–1993)	9.2[f]	7.6 (1993)[f]	17 (1985–1993)

[a] Original Federal States only.

[b] Original and New Federal States.

[c] Per hectare of land used (crops and grassland) without alpine pastures, para-agriculture enclosed.

[d] Figures for 1993.

[e] Crops and grassland.

[f] All agricultural land.

Source: OSPAR, 1995

been achieved, however, do not match the goal of a 50% reduction by 1995, which was the original objective.

The theoretical effect of measures A1–A5 of the Code of good agricultural practice of the Nitrate Directive, assuming that these measures to reduce run-off are implemented, show that for the eastern and southern parts of the country, these effects could be considerable, up to 100 mg/l reduction in nitrate in surface water (Fig. 3.7; de Cooman *et al.*, 1995).

The theoretical effects of measure A6 of the Code on the quality of surface waters, shallow- and deep groundwater, assuming that limits on application of manure, mineral fertilizer and equilibrium fertilization are respected, are shown in Figures 3.8–3.10 (de Cooman *et al.*, 1995). These figures show that both reduction of manure application and equilibrium fertilisation could have a major impact on the different water qualities in the Netherlands, especially in the southern and eastern parts of the country.

ROLE OF DIFFERENT ACTORS

The actors that influence the policy process can be divided into three categories, the government, the agricultural lobby and the environmental organizations.

Table 3.16 Inputs of nutrients from agriculture and expected results of the meassures taken or planned.

	1985 (tot-P) (tonnes)	1992 (tot-P) (tonnes)	1995 (tot-P) (tonnes)	% reduction 1985–1992	% reduction 1985–1995
Phosphorus					
Belgium	2 470	1 680	<1680	32	>32
Denmark	560	560	530	0	5
France[a]	25 400	24 000	21000	6	17
Germany (Original Federal States)	17 100	17 100[b]	13500	0[b]	21
Netherlands[c, d]	6 400	6700	6400	−5	0
Norway[e]	290	246	202	15	30
Sweden[f]	390	370	270	5	31
Switzerland[g]	408	335[h]	18[h]	25[h]	
United Kingdom	NI	NI	NI	NI	NI
Nitrogen					
Belgium	39 580	35 350	<35 350	11	>11
Denmark	59 000	59 000	50 000	0	15
France[a]	200 000	195 000	180 000	3	10
Germany (Original Federal States)	324 000	324 000[b]	270 000	0[b]	17
Netherlands[c,d]	137 000	134 000	116 000	2	15
Norway[e]	12 640	11 406	9 827	10	22
Sweden[f]	21 000	17 000	15 000	19	28
Switzerland[g]	10 800	9 400[i]	8 700[i]	13[i]	19[i]
United Kingdom	NI	NI	NI	NI	NI

[a] Provisional estimates.

[b] Results for 1990, respectively for the period 1985–1990.

[c] Measured as PO4-P.

[d] Provisional figures. Figures for 1995 results from model calculations to Dutch surface waters based on measures taken. They do not include all measures under concideration. The model takes into account the retarded effect due to retention resulting in higher reductions at a later stage.

[e] Figures relating to identified problem areas.

[f] Refers to the catchment area. The figures are uncertain and are based on estimates. The timetable of the results of the measure is provisional.

[g] figures refer to nutrient input to waters of the Rhine watershed downstream of the lakes inside Switzerland calculated for the year 1986 (the national background load of 56 tonnes for P and 6445 tonnes for N is not included in the figures).

[h] Rough estimates (the reduction potential for 2005 is estimated by 51%).

[i] Rough estimates (the reduction potential for 2010 is estimated to be 43%).

NI No information provided as UK is not subject to the 50% reduction commitment.

Source: OSPAR 1995

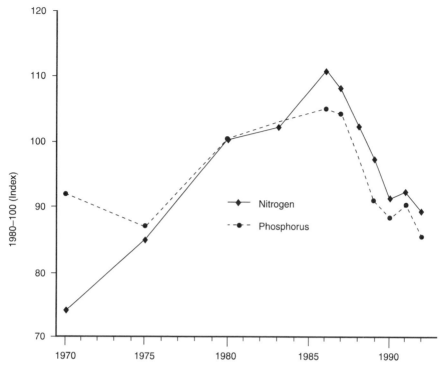

Figure 3.6 Development of mineral surpluses in agriculture 1970–1992. Source: CBS

Within the government, three different Ministries are involved. First, the Ministry of Agriculture, Nature Management and Fisheries is responsible for the agricultural sector, most specifically manure and ammonia policies. Second, the Ministry of Housing, Spatial Planning and the Environment (VROM) is responsible for drinking water quality and policies concerning the atmosphere. Third, the Ministry of Transport, Public Works and Water Management is responsible for monitoring and maintaining water quality standards for surface- and groundwater, both fresh water and the North Sea.

The agricultural lobby in the Netherlands is well organized and maintains close ties with the Ministry of Agriculture. It carries a lot of weight, because considerable economic interests are at stake: agriculture contributes 5% to the Gross National Product (GNP) of the Netherlands.

Environmental organizations such as Greenpeace, 'Natuur en Milieu' and the Centre for Agriculture and Environment, try to influence policy by informing the public and influencing public opinion through the media. 'Natuur en Milieu' tries to affect policy by maintaining close contacts with civil servants and advisory councils.

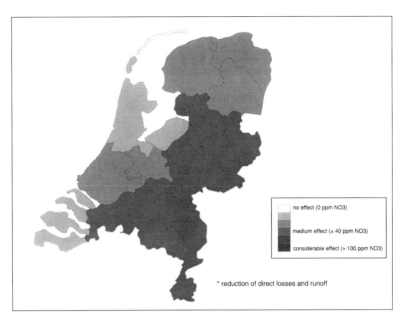

Figure 3.7 Theoretical effect of measures A1–A5 on the quality of surface water. Source: de Cooman *et al.*, 1995

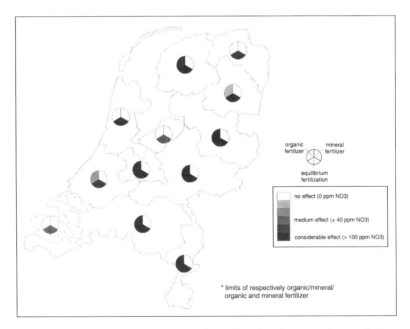

Figure 3.8 Theoretical effect of measure A6 on the quality of surface water. Source: de Cooman *et al.*, 1995

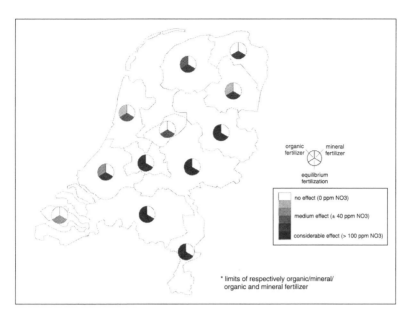

Figure 3.9 Theoretical effect of measure A6 on the quality of shallow groundwater. Source: de Cooman *et al.*, 1995

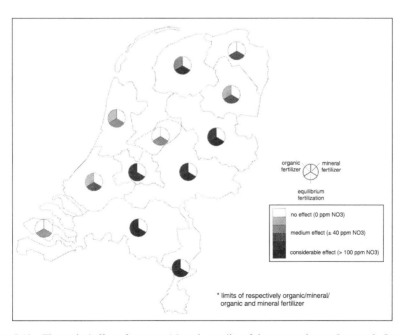

Figure 3.10 Theoretical effect of measure 16 on the quality of deep groundwater. Source: de Cooman *et al.*, 1995

FUTURE DEVELOPMENTS

In February 1997, policy was decided for the next 10 years, with the stipulation that there will be an evaluation in two years (2000), after which adjustments can be made. The proposed measures have so far met with a lot of resistance from farmers' organizations. The measures that have been decided on are, therefore, less stringent than the original proposals, in order to facilitate the acceptance of the new MINAS system by farmers. Consequently, the levies that will have to be paid between 1998 and 2000 will be low. It is to be expected that during that period there will be no great reductions in mineral production and utilization in agriculture. However, during this period, a number of important adjustments will have to be made, that will hopefully create the preconditions for more substantial reductions in the future.

The introduction of the Minerals Accounting System (MINAS) in 1998 for qualifying farms will involve registration of the mineral inputs (nitrogen and phosphate) used on the farm in animal feeds and mineral fertilizer, and the mineral output in the form of products and manure. It is also meant to be a management tool, which will give farmers more insight into and control over the mineral economy of their operation.

Farmers have a choice between an exact declaration and a standard amount. The specific method is intended for those farmers who can significantly influence mineral contents in inputs and outputs. It involves more record keeping and also means that all manure leaving the farm has to be sampled and tested for mineral content. The standard amounts are more stringent and will result in larger levies, which will create an incentive to use the exact declaration. The intention is to extend the MINAS system to all farms in the future.

During the next two years, the calculation of phosphate losses will exclude phosphates from mineral fertilizer; after 2000, phosphates from both manure and fertilizer will be included. The acceptable loss norms will also be lowered, which means that after 2000 norms will be lowered and levies will go up, incentives for more substantial reductions.

Experiments are being considered with integrated environmental licences, in order to reduce the burden of regulation, both on farmers and on the government. In selected municipalities, manure and ammonia regulations may be temporarily lifted and replaced by environmental licences. Farmers with a good environmental management record, would receive such a licence, which covers ammonia emissions, manure distribution, mineral ledger and pesticide use.

Due to the exceptional intensity of Dutch livestock production compared with other EU member states, these policy intentions are at odds with the EU Nitrate Directive. In the Dutch situation, the implementation of the Directive requires drastic measures which take time. Also, the goals will have to be met by a different route, which will require consultations with the European Commission. According to present data, loss standards of 180 to 200 kg N/ha are required to meet the objective from the Rhine and North Sea Action Programme to reduce the nitrogen level by 50% relative to 1985.

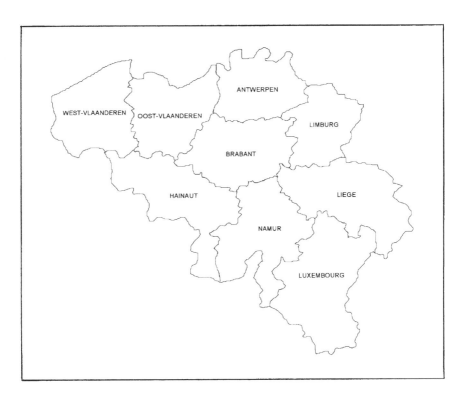

BELGIUM

CHAPTER 4: BELGIUM

COUNTRY CONDITION

The total area of cultivated land in Belgium occupies about 1.4 million ha; this area represents about 45% of the total territory of the country, which is around 3 million ha. The agricultural sector consists of about 93 000 holdings and employs more than 100 000 people. Between 1980 and 1990 the number of holdings decreased by about 20% and at the same time, the average size of farms increased from 14 to 15 ha. The total value of agricultural production is 6 119 million ECU (de Cooman *et al.* 1995).

Of the total agricultural area, 44% is situated in Flanders, 46% in Wallonia and 10% in the Brussels area. In the Flemish region, farms are generally small: 12 ha on the average. When small farms (less than 2 ha or <300 kg P_2O_5 production from animal manure per year) are not considered, the average size of Flemish farms is 14 ha. Traditionally, these were small mixed farms, but gradually they have become very specialized, intensive livestock farms. About 43% of the agricultural area in Flanders is grassland, the remaining 57% is arable land. On the sandy soils in the northern part of Flanders, fodder becomes important: in the province of Antwerp, fodder is grown on about 70% of the arable land. The middle part of the country, which is part Flanders, part Wallonia, has a rich silty soil, where cereals are grown (Baldock and Bennett, 1991).

In Wallonia, the average farm size is about 25 ha; these holdings are mostly cropping and dairy farms. Of the total agricultural area, 42% is grassland; in Wallonia about half of the total area is grassland. Sugar beets are an important crop and cover about 20% of arable land in the provinces of Hainaut and Liège (Wallonia) and Brabant.

The livestock population of Belgium totals about 4.2 million Livestock Units (LU), which is about 4% of the European livestock population. It consists of 54% cattle, 38% pigs and 7% poultry. Between 1983 and 1990, the livestock population continued to grow, but it has stabilized since 1990. The structure of the livestock population in Flanders is very different from Wallonia (Table 4.1).

As Table 4.1 shows, cattle farming is much more prominent in Wallonia, while pig and poultry farming is much more important in Flanders. The livestock density in the province of Antwerp is the highest in the country and amounts to 7.7 LU/ha, while in the other Flemish provinces it is between 5 and 7 LU/ha. In Wallonia, however, it is not more than 2 LU/ha.

51

ENVIRONMENTAL PRESSURES AND SOURCES

Use of mineral fertilizer

Use of mineral fertilizer in Belgium is fairly high compared with that in other European countries. Nationally, the average use is 163 kg N/ha, but the average for Flanders is 192 kg N/ha, whereas the average for Wallonia is only 136 kg N/ha. These figures are based on calculations by de Cooman *et al.* (1995), using 1990 REGIO-banque data. More recent data from the Vlaamse Mestbank show an average of 86 kg N/ha and 20 kg P_2O_5 for Flanders.

As Table 4.2 shows, the total utilization of mineral fertilizer has increased from 1970 until 1986 and has since decreased somewhat. The use of mineral fertilizer in the different provinces is shown in Table 4.3. Recent data for Flanders (1994) from another source (Vlaamse Mestbank) show lower average use figures per ha. It is not clear whether these lower figures reflect a genuine decrease in mineral fertilizer use or if they are mainly due to a different method of data collection (Table 4.4).

Use of animal manure

The average quantity of manure produced in Flanders is 300 kg N/ha, while in Wallonia it is 145 kg N/ha. Table 4.5 shows the averages for the different provinces, with Brabant being the only province in Flanders that produces less than the limit, like the provinces in Wallonia do, except for Hainaut (Figure 3.1). The other Flemish provinces all produce considerably more than the limit.

More recent data for Flanders (Table 4.6), collected by the Vlaamse Mestbank over 1995 in an inventory of all holdings over 2 ha and a phosphate production of over 300, show average organic fertilizer production and utilization per ha, for the Flemish provinces. The origin of differences in Tables 4.5 and 4.6 are not clear.

In Table 4.7 the quantitites of mineral fertilizer and organic manure have been added for each province and are compared with equilibrium fertilization limits. The results show that two provinces in Wallonia stay well under the limit, as does Brabant; the two others slightly exceed the limit. For Flanders, however, it is clear that all provinces except Brabant, exceed the limit considerably, notably West-Vlaanderen and Antwerp. It should be noted, however, that some of the manure produced in the Flemish provinces is transported to other regions, so not all of it is applied locally (see below).

Nitrogen balance

The average nitrogen balance for the country as a whole is 178 kg N/ha, but there are large differences between the different provinces (see Table 1.3). The highest nitrogen balance has the province of Antwerpen (358 kg N/ha), followed by West- and Oost-Vlaanderen (308, resp. 286 kg N/ha). The nitrogen balance for Wallonia is much lower, 119 kg N/ha for Liege, 136 kg N/ha for Hainaut (Schleef and Kleinhauss, 1995).

Table 4.1 Structure of the livestock population

NUTS II	Total 1000 LU	Cattle (%)	Pigs (%)	Poultry (%)
Wallonia				
Namur	211	95.7	2.8	0.5
Hainaut	330	89.4	8.2	1.5
Liege	288	88.5	9.7	1.0
Luxembourg	277	97.5	1.4	0.4
Flanders				
Brabant	233	68.2	24.5	5.6
Limburg	332	40.1	43.1	16.3
Oost-Vlaanderen	705	45.8	45.7	7.8
West-Vlaanderen	1282	30.1	62.6	6.8
Antwerpen	570	42.5	38.4	18.4

Source: de Cooman *et al.*, 1995

Table 4.2 Utilization of mineral fertilizer in Belgium and Luxembourg

Year	1970	1975	1980	1982	1984	1986	1988	1990
Utilization (1000t)	178	182	194	197	199	199	196	186

Phosphorus

The total mineral production from livestock amounted to 32 million kg phosphorus, mostly originating from cattle and pigs. The mean use of P from farm yard manure is 54 kg P/ha and from mineral fertilizer 25 kg P/ha; the amount of P from mineral fertilizer has been stable since 1950 (see Tables 4.4 and 4.6).

AMBIENT ENVIRONMENTAL CONDITIONS OF SOILS, SURFACE-, GROUND- AND DRINKING WATER

Surface water

Water quality of surface waters in Wallonia is good or very good in 50% of the sampling points, mediocre in 15%, in 20% of the sampling points it is bad and in the last 20%, it has really deteriorated. Eutrophication in Wallonia is fairly widespread but becomes less as one goes further south.

In Flanders, almost all surface waters show some degree of eutrophication. One research study concluded that in 1991 about 75% of all surface water could be considered biologically dead (Vosmer, 1995). Figures 4.1 and 4.2 show the concentrations of NH_4-N and NO_3-N at different sampling points in Flanders. In the province of West-Vlaanderen, surface water is of particularly poor quality and levels of N above 40 mg/l are not uncommon. Another factor that contributes to the poor quality of surface water in Belgium is the relative absence of adequate

Table 4.3 Utilization of mineral fertilizer (kg N/ha/year), compared with limits based on equilibrium fertilization

NUTS II	Limit	Utilization (1990)	Difference
Wallonia			
Luxembourg	154.3	82.5	71.8
Namur	163.7	125.5	38.2
Liege	158.1	151.0	7.1
Hainaut	167.5	161.7	5.8
Flanders			
Limburg	169.2	175.0	5.8
Brabant	161.2	176.0	14.8
Antwerpen	177.3	203.4	26.1
West-Vlaanderen	168.7	199.4	30.7
Oost-Vlaanderen	170.1	202.7	32.6

Source: De Cooman *et al.*, 1995

water treatment facilities, especially in Wallonia and Brussels. The region of Brussels drains the wastewater of about one million people untreated. In Flanders, only 54% of all wastewater is treated (Vosmer, 1995).

Groundwater
In Wallonia, most drinking water is of good quality. As Table 4.8 shows, the EU standard for drinking water of 50 mg NO_3/l is exceeded only rarely. In Flanders, the norm of 50 mg nitrate/l is exceeded in 28 municipalities or about 2% of the total drinking water supplies. High levels of nitrate in groundwater do not always correspond to areas with high fertilization levels; instead, it seems more related to the nature of the soil and the characteristics of deeper ground layers (Figure 4.3).

In Belgium, 80% of the drinking water supply comes from groundwater.

Table 4.4 Utilization of mineral fertilizer in Flanders, based on data supplied to the Vlaamse Mestbank, 1994

Province	Nitrogen (kg N/ha)	Phosphate (kg P_2O_5/ha)
Flanders	86	20
Limburg	70	21
Brabant	108	38
Antwerpen	55	13
Oost-Vlaanderen	96	25
West-Vlaanderen	90	12

Source: Annual Report Vlaamse Mestbank, 1995

Table 4.5 Production of organic fertilizer (in kg N/ha x year), compared with limits based on equilibrium fertilization

NUTS II	Limits	Production	Difference
Wallonia			
Namur	128,7	117,6	–11,1
Luxembourg	177,5	175,3	–2,2
Liege	152,2	152,2	0,0
Hainaut	112,6	131,0	18,4
Flanders			
Limburg	200,7	273,6	72,9
Oost-Vlaanderen	202,2	317,7	115,5
West-Vlaanderen	199,9	360,9	161,0
Antwerpen	213,7	426,5	212,8

Source: de Cooman *et al.*, 1995

Table 4.6 Production and utilization of organic fertilizer, flemish provinces

Provinces	Production (kg N/ha)	Production (kg P_2O_5/ha)	Utilization (kg N/ha)	Utilization (kg P_2O_5/ha)
Flanders	251	113	247	109
Brahant	124	50	131	55
Limburg	200	91	226	111
Oost-Vlaanderen	246	105	248	105
West-Vlaanderen	297	136	282	124
Antwerpen	312	146	288	125

Source: Vlaamse Mestbank, 1994

Wallonia is the main supplier of drinking water and exports about 40% of its own production to Flanders. About 5–10% of the Belgian population depends on private wells for drinking water and their quality is often doubtful: in Flanders, 40% of private wells have nitrate levels that exceed the standard, 17% have phosphate levels over 0.4 mg/l. For the 5–10% of the population who are not connected to the public water supply, these high levels pose a serious risk (Table 4.9).

Lakes and marine waters

There are very few lakes in Belgium and no data are available as to their water quality. Marine waters up to 30 km from the Belgian coast have elevated N-levels of more than 0.21 mg N-total/l (see Figure 3.4).

Table 4.7 Mineral and organic fertilizer (kg N/ha/year)

NUTS II	Limit	Utilization	Difference
Wallonia			
Luxembourg	324.6	257.8	−66.9
Namur	271.3	243.1	−28.2
Liege	297.7	303.2	5.5
Hainaut	253.3	297.7	39.4
Flanders			
Brabant	307.5	296.4	−11.1
Limburg	330.7	448.6	117.9
Oost-Vlaanderen	333.3	520.4	187.1
West-Vlaanderen	329.2	560.3	231.1
Antwerpen	345.3	529.9	275.7

Source: de Cooman *et al.*, 1995

NEED FOR AND PURPOSE OF EXISTING POLICY AND REGULATION (INTERNATIONAL AND NATIONAL)

International policy

Belgium has been a participant in all INSC and PARCOM conferences and as an EU member, is responsible for implementing the Nitrate Directive. Flanders has designated its whole area as vulnerable zone, while Wallonia has designated two separate zones. As of June 1995, a Code of good agricultural practice had not yet been notified, but this may have happened in the meantime, since the new Manure

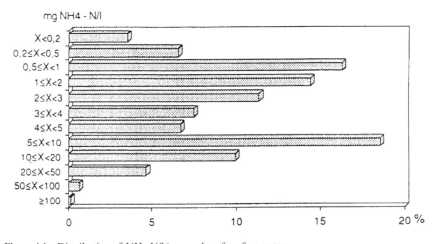

Figure 4.1 Distribution of NH_4-N/l in samples of surface water

mg NO3 - N/l

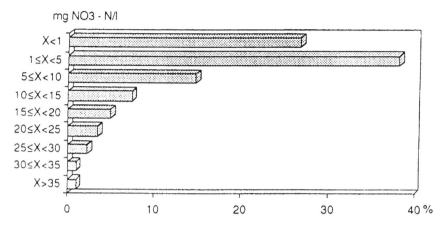

Figure 4.2 Distribution of NO₃-N/l in samples of surface water

Table 4.8 Concentration of nitrates on groundwater, 1990

Aquifers by soil type	Number of wells	<10mg NO₃/l	10 – 25 mg NO₃/l	25 – 50 mg NO₃/l	>50 mg NO₃/l
Limestone	73	45%	36%	19%	0%
Chalk	38	42%	42%	16%	0%
Sand	46	50%	13%	28%	9%
Alluvial	22	50%	33%	4%	4%
Total	179	47%	31%	19%	3%

Decree for Flanders, which encompasses the legislation required by the Nitrate Directive, has become operable on 1 January 1996.

National/regional policy

Belgium is a federal state composed of three regions, namely Flanders, Wallonia and Brussels, where agriculture is of little importance. In Belgium, agriculture is partially a federal and partially a regional responsibility, falling within the competence of the Ministry of Agriculture and the Ministry of the Economy respectively. Environmental protection, however, is a regional competence, and each region is responsible for its own administration and legislation. Because nutrient pollution through agriculture is mainly a problem in Flanders, we will focus on policy and legislation in that region.

As will be apparent from the data presented, the main problems in Belgium are caused by the intensive livestock farming in Flanders and the resultant manure surplus. The most serious problem is the contamination of groundwater by nitrates, but the eutrophication of surface waters, particularly the Schelde and Jzer basins, are also cause for concern. Belgium has initially responded slowly to existing

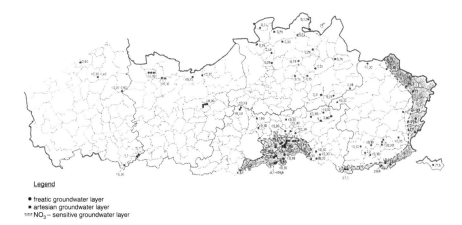

Legend

● freatic groundwater layer
■ artesian groundwater layer
▬ NO₃ – sensitive groundwater layer

Figure 4.3   Water quality in the most important sources

Table 4.9   Nitrate and phosphate in well-waters in Flanders (1988–1989)

Province	Number of wells sampled	Nitrate level (mg/l)			Phosphate level (mg/l)	
		>25	>50	>100	>0.4	>5
Antwerp	92	32	19	7	16	2
		(34.78%)	(20.65%)	(7.61%)	(17.39%)	(2.17%)
Brabant	281	142	95	41	37	4
		(50.53%)	(33.81%)	(14.59%)	(13.17%)	(1.42%)
Limburg	333	160	105	44	12	0
		(48.05%)	(31.53%)	(13.21%)	(3.60%)	(0.00%)
East Flanders	576	341	269	136	126	13
		(59.20%)	(46.70%)	(23.61%)	(21.88%)	(2.26%)
West Flanders	292	117	89	58	77	10
		(40.07%)	(30.48%)	(19.86%)	(26.37%)	(3.42%)
TOTAL	1574	792	577	286	268	29
		(50.32%)	(36.66%)	(18.17%)	(17.03%)	(1.84%)

Source: Milieubeleidsplan en Natuurontwikkelingsplan voor Vlaanderen

problems. Coordinating legislation across different levels of government has slowed down the process.

The measures that were in effect until 1996 have had insufficient effects on limiting agricultural pollution. The Environment and Nature Plan (MINA Plan 2000) for Flanders, outlined environmental policies for 1990–1995. It consisted of two parts: an inventory of the environmental situation, and a number of action plans. The objectives were to reduce ammonia emissions by 20% by the year 2000, on a 1988 baseline. Objectives for groundwater quality were for nitrates 35 mg/l, for phosphates 2 mg/l in the year 2000. In 1991, the Manure Decree was accepted, utilizing fairly liberal limits. The Manure Action Plan, that was developed in the context of MINA, has led to a tightening of these standards and a number of measures, that are much stricter and more limiting. These measures have become part of the revised Manure Decree, which went into effect in 1996. The new policy contains a lot of elements that are similar to Dutch manure policies and has also been influenced by Danish manure legislation.

The new MINA Plan 1997–2002 aims for an equilibrium in input and output of N, P and K, in which standards are more closely related to the carrying capacity of the ecosystem. To this end, an evaluation of fertilization standards by 1/1/1999 is planned. Agricultural and environmental policies will aim for similar objectives.

POLICY FORMULATION AND LEGAL INCORPORATION

Laws in Belgium that aim to protect the environment from nutrient pollution are the following:

1. Law on the protection of surface waters (1971), supplemented by a 'royal decree' (1976) and a decree of 1985, to implement the law. It encompasses measures for the protection of minimal water quality and some quality objectives for waters for specific purposes.
2. Wastewater treatment plans, that are still being implemented, to upgrade treatment facilities to European standards.
3. Regional decree for the protection of groundwater (1984) limits the application of nitrogen to agricultural land to 400 kg N/ha/year, in combination with groundwater quality objectives.
4. In 1989, a Treaty was signed between the federal government and the regions. In this treaty, the Ministers concerned agreed to aim at a substantial reduction of toxic substances and nutrients to the North Sea. This treaty sets no specific objectives and does not name the specific substances that should be reduced.
5. The Manure and Fertilizer Decree (1991). This decree regulates not only animal manure, but all fertilizers and other substances that can supply nutrients to the environment. It sets fertilizing limits for both N (400 kg/ha/year) and P_2O_5 (200 kg/ha/year), defines mineral surplus, sets a levy on mineral surplus and period when spreading manure is prohibited. It also organizes the distribution of manure, through additional measures. Disposal of manure surpluses have to be documented. Since fertilizing limits were fairly high, surplus problems could be solved by distribution from surplus farms to other regions.
6. The Manure Action Plan (1993). The objective of this plan is to reach a sustainable nutrient balance in Flanders and the implementation of the EU Nitrate Directive. It proposed stricter fertilizing rules, protection of vulnerable areas, the regulation of farm structure and a levy system. Furthermore, in order to

bend the trend of growing manure surpluses, a stand-still on manure production at the 1992 level is agreed upon. At farm level, this means that further increase of the number of livestock is bound to strict rules: farmers can only expand if others reduce their livestock population.

The discussion that followed the approval of the Manure Action Plan took place between farmers' organizations, who were shocked by the proposals and the environmental lobby, which was very much in favour of it. During these discussions, the opposing organizations found an issue that they could both agree upon, some 'common ground', which was the value of the 'family livestock farm'. Their common 'enemy' was the industrialized farms, who were cast in the role of being the main producers of the manure surplus and thus the greatest polluters. Thus industrialized farms were blamed for the severity of the manure problem, while the 'family livestock farm' was portrayed as the provider of food, the backbone of the rural community and the defender of moral and Christian values. It also became the defender of ecological principles in agriculture, as opposed to the purely commercial interests pursued by the industrial farms. On the basis of this common ground, a compromise between opposing camps was reached and the positive discrimination of the 'family livestock farm' was proposed. A legal definition of the family livestock farms was worked out and special measures concerning the transport of surplus animal wastes were agreed upon, to favour these kind of farms (Van Gijseghem, 1996).

7. The New Manure and Fertilizer Decree (1996). This decree is the legal incorporation of the Manure Action Plan. It sets new fertilizing standards for the years 1996–2003, which set stricter limits on both 'use of all fertilizer' and on 'use of mineral fertilizer', and sets a levy. It requires a manure accounting, which accounts for the production, application and distribution of manure and fertilizer. It also sets specific limits for protection areas, of which there are three kinds: the 'group nature', the 'group water' (protection of ground- and surface water) and the phosphate saturated areas. It also introduces the concept of the family livestock farm, which is legally defined. Family livestock farms are positively discriminated in trading and transporting their manure surpluses.

The Manure and Fertilizer Decree also prohibits the application of manure and mineral fertilizer during certain periods and regulates the operations of the Manure Bank, transport and distribution of manure, record keeping and the positive discrimination of the family livestock farm in this. It also enforces the 'stand-still' of the total livestock population, through the municipal permit system.

REGULATORY CONTROLS

In the Manure and Fertilizer decree of 1996, the following measures have been included.

Definition of the Family Livestock Farm

The special status is established as a 'family livestock farm', which is based on the following definition:

- The farmer must be owner or tenant of the farm, he and his family must work there full time, only one employee is accepted, and he must take all economic decisions himself.
- The size of the livestock: up to 1800 pigs or 100 dairy cows, 300 other cows or 600 calves or 70 000 laying hens.
- There may be no bonds between the providers of feed, young animals and the purchasers of the fullgrown animals.
- The family livestock farm must have enough land to use 25% of the amount of manure that is produced at the farm.

Each year, the farmer has to ask to be notified as a family livestock farm and has to supply documentation to support his claim. The 'positive discrimination' of family livestock farms allows them to trade their surplus manure with neighbours or farms in neighbouring communities, while non-family livestock farms are obliged to transport their manure surpluses over greater distances to regions without a surplus. Family livestock farms also have different fertilizing limits in some vulnerable areas.

Manure accounting

- All holdings that have produced at least 300 kg phosphate in animal manure or who cultivate more than 2 ha of agricultural land, have to submit a manure accounting form.
- All transporters, exporters, processors of manure and managers of manure collection points also have to submit manure accounting.
- Producers of other fertilizers like compost, sludge etc. have to submit an accounting.
- All producers of animal manure or other organic waste, who have produced more than 2000 kg phosphate, are required to maintain a register and keep a file of all pertinent documents. This is also required of all transporters and traders of manure.

Fertilizing limits

There are four different kinds of fertilization limits, depending on the location of the holding: general limits for all agricultural land outside the protection areas, and limits for three groups of protection areas: the water (ground- and surface water protection), the nature and the phosphate-saturated areas. Protection areas are designated on small scale maps, that are supplied to all farmers. Larger scale maps can be consulted at the local municipalities. The general fertilization limits are given in Table 4.10. Limits for the water group are given in Table 4.11, for nature group in Table 4.12 and for phosphate-saturated areas in Table 4.13. In the

Table 4.10 Fertilization limits 1996–1998 (kg/ha per year).

| Crop | Phosphate (P_2O_5) | Nitrogen (N) | |
	All fertilizers (= kg P_2O_5)	All fertilizers (= kg N total)	Mineral fertilizers (= kg mineral N)
Grassland	170	450	250
Maize			
1996	160	325	200
1997	155	325	200
1998	150	325	200
Crops with low N-requirement	125	170	125
Other crops	150	325	225

Source: VLM, 1995

Table 4.11 Fertilization limits in the 'group nature' (kg/ha per year)

| Crop | Phosphate (P_2O_5) | Nitrogen (N) | |
	All fertilizers (= kg P_2O_5)	All fertilizers (= kg N total)	Mineral fertilizers (= kg mineral N)
Grassland	120	400	200
Maize	100	275	150
Crops with low N-requirements	80	125	70
Other crops	100	275	150

Source: VLM, 1995

phosphate-saturated areas, less stringent limits are set for family livestock farms than for other holdings.

Application of manure and fertilizers

General rules for the application of manure and other organic fertilizer prohibit the application of manure and other organic fertilizer during the following periods:

- from 21 September until 21 January;
- on all Sundays and holidays;
- on all Saturdays except for the period between 1 February and 15 May;
- at night between 10 p.m. and 7 a.m.;
- when the soil is frozen, inundated or covered with snow.

Special rules apply for protection areas (Fig. 4.4).

Manure surpluses and transport of manure

The manure bank calculates whether a holding has a manure surplus or not, based on a simple formula: manure surplus = manure production + use mineral

Table 4.12 Fertilization limits in 'group water' (kg/ha per year)

Crop	Phosphate (P_2O_5)	Nitrogen (N)		
	All fertilizers (= kg P_2O)	All fertilizers (= kg N total)	Organic manure	Mineral fertilizers (= kg mineral N)
Grassland	120	350	200	200
Maize	100	275	170	150
Crops with low N-requirements	125	170	125	125
Other crops	150	325	170	225

Source: VLM, 1995

Table 4.13 Fertilization limits for P-saturated areas (kg/ha per year), differentiated for family- and non-family livestock farms.

Farm type	Chemical P_2O_5 (kg/ha)	Organic P_2O_5 (own manure only, own soil)	Total
Non-family livestock farm	40	0	40
Family livestock farm	40	grassland : 80 other crops : 60	grassland : 80 other : 60

Source: VLM, 1995

fertilizer + use other fertilizer – maximum allowable use of manure on land belonging to the farm. For manure production figures, standard amounts are used. Manure surpluses can be distributed to other holdings according to strict rules. Generally speaking, surplus manure has to be transported by licensed transporters, unless family livestock farms transport manure to neighbouring farms or farms in neighbouring communities. All transfers have to be documented.

Non-family farms and large holdings (phosphate production > 10 000 kg) situated in areas with high levels of manure production, are obliged to transport their manure surpluses to areas with low levels of manure production. From 1999 all large holdings, as defined above, will be obliged to either export their manure surpluses or to process them. These transports can only be executed by licensed manure transporters and require transport documents. Table 4.14 shows how much manure has been transported in the context of family livestock farms transporting manure (in kg N) to neighbouring farms or neighbouring communities in 1995. Table 4.15 shows manure transports with transport documents (in kg N) in 1995.

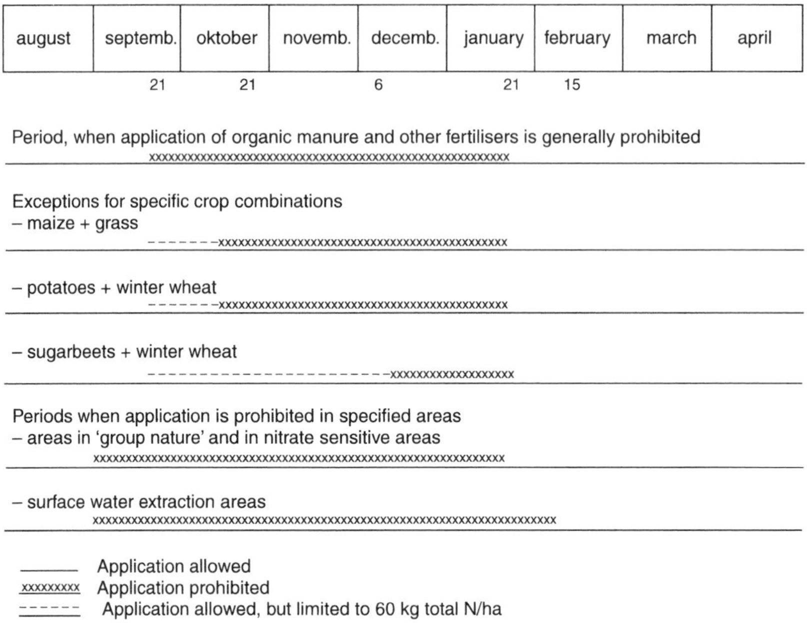

august	septemb.	oktober	novemb.	decemb.	january	february	march	april
	21	21		6	21	15		

Period, when application of organic manure and other fertilisers is generally prohibited
xxxxxxxxxxxxxxxxxxxxxxxxxxxxxxxxxxxxxxxxxxxxxxxxxxxxxxx

Exceptions for specific crop combinations
– maize + grass
– – – – – –xxxxxxxxxxxxxxxxxxxxxxxxxxxxxxxxxxxxxxxxxxxx

– potatoes + winter wheat
– – – – – –xxxxxxxxxxxxxxxxxxxxxxxxxxxxxxxxxxxxxxxxxxx

– sugarbeets + winter wheat
– – – – – – – – – – – – – – – – – –xxxxxxxxxxxxxxxxxxx

Periods when application is prohibited in specified areas
– areas in 'group nature' and in nitrate sensitive areas
xxxxxxxxxxxxxxxxxxxxxxxxxxxxxxxxxxxxxxxxxxxxxxxxxxxxxxxxxx

– surface water extraction areas
xxxxxxxxxxxxxxxxxxxxxxxxxxxxxxxxxxxxxxxxxxxxxxxxxxxxxxxxxxxxxx

——— Application allowed
xxxxxxxx Application prohibited
– – – – – Application allowed, but limited to 60 kg total N/ha

Figure 4.4 Periods when application of manure and fertilizer is prohibited

Table 4.14 Manure transport in kg N to neighbouring farms or communinities

Destination Origin	Antwerp	Fl. Brabant	W. Flanders	E. Flanders	Limburg	Total
Antwerp	3 847 540	30 909	2 796	4 685	34 297	3 920 227
Fl. Brabant	29 236	591 784	0	11 578	58 010	690 608
W. Flanders	4 786	956	8 290 398	127 869	0	8 424 009
E. Flanders	51 273	16 980	80 669	3 970 043	0	4 118 965
Limburg	8 437	41 071	0	0	1 865 483	1 914 991
Total	3 941 272	681 700	8 373 863	4 114 175	1 957 790	19 068 800

Source: VLM, 1995

Stand-still of phosphate production and permits

In 1992, the decision was made that the total quantity of phosphate and nitrogen produced as livestock manure should be brought to a stand-still: the 1992 quantity, taken as the production ceiling, was calculated at 75 million kg phosphate and 169 million kg nitrogen. This means that the total livestock population cannot expand beyond 1992 levels. The possibilities for expansion are determined per municipality: all municipalities in Flanders are categorized in one of four classes: white to pale-grey, dark-grey and black (Fig. 4.5). In white and grey municipalities, there are some possibilities of expanding,

Table 4.15 Manure transports with transport documents to destinations in Flanders, from different origins (upper half) and from origins in Flanders to destinations outside Flanders (lower half)

Origin \ Destination	Antwerp	Fl. Brabant	W-Flanders	E-Flanders	Limburg	Total Origin
Antwerp	3 118 163	706 671	45 533	53 741	805 319	4 783 427
Flemish-Brabant	13 339	191 530	37 717	36 597	11 398	290 581
West-Flanders	50 135	468 305	6 505 589	1 686 541	6 985	8 717 555
East-Flanders	42 090	90 739	417 128	2 084 817	68 247	2 703 021
Limburg	36 932	87 231	0	0	1 041 444	1 165 607
Brussels	4 415	2 340	13 912	1 020	61 464	83 151
Wallonia	314	850	33 949	13 377	8 291	56 781
Netherlands	37 775	0	28 524	4 164	1 190 347	1 260 810
France	0	0	178	0	0	178
Germany	0	0	0	0	532 694	532 694
Total destination	3 303 163	1 601 666	7 082 530	3 880 257	3 726 189	19 663 805

Origin \ Destination	Brussels	Wallonia	Netherlands	France	Germany	Total Origin
Antwerp	91 925	367 239	313 325	109 884	0	882 373
Flemish-Brabant	154	23 171	1 050	0	0	24 375
West-Flanders	0	213 711	453 601	1 257 119	0	1 924 431
East-Flanders	4680	41 311	322 789	107 997	0	472 097
Limburg	0	25 407	3 485	6 372	10 743	46 007
Total destination	96 759	670 839	1 094 250	1 481 372	10 743	3 353 953
Grand total destination/origin						22 947 768

Source: VLM, 1995

depending on strict conditions. In dark-grey and black municipalities these possibilities are very limited: only if the livestock population here has decreased to create a certain 'reserve' can other farms expand with 75% (dark-grey) or 50% (black) of the existing reserve. This means that a certain percentage of the existing rights to hold livestock is 'siphoned off' when farms are sold or terminated, comparable to the manure production rights in the Netherlands. The permit policy has become an important instrument in keeping manure production in Belgium within bounds and the municipalities play an important role in the implementation of this policy.

Code of Good Agricultural Practice
The requirements for the Code of Good Agricultural Practice, in order to meet the EU Nitrate Directive, are being met through the new Manure Decree.

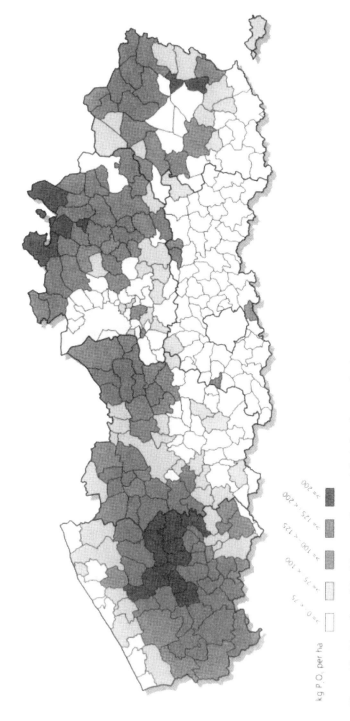

Figure 4.5 Map of municipalities, according to original phosphate production pressure

MONITORING AND CONTROL, RECORD KEEPING AND VERIFICATION

Implementation of the Manure Decree

Control of Manure Decree measures is the responsibility of different officials: civil servants of the manure bank, environmental inspection and federal police; municipal police can check rules concerning application of animal wastes. Submitting the farms' manure accounts, not paying levies, ignoring application limits etc., can all lead to fines or even jail sentences. These control measures have been in force since 1991.

FINANCIAL MEASURES AND INCENTIVES

The new Manure Decree has different levies.

The basic levy
This is the most comprehensive levy and must be paid by all holdings with a weighed phosphate production of more than 1500 kg. The government takes the

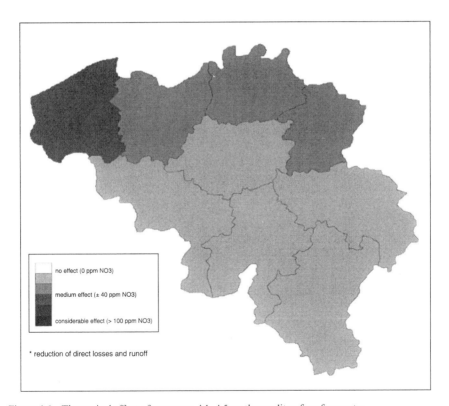

Figure 4.6 Theoretical effect of measures A1–A5 on the quality of surface water

position that every kilo of minerals adds to the manure problems and thus all farms, even ones with no manure surplus, should pay this levy.

A graduated system of charges has been worked out on the principle that the more manure is produced, the higher the levy should be. The system has four different brackets (Table 4.16). Manure production in this table is weighed: it includes all animal manure, except that from laying hens, plus 40% of the laying hen manure. All manure that is exported or processed will be charged in the first bracket; in fact, the quantities of manure exported etc., should be added on to the limit of the first bracket. This means that a levy has to be paid for all manure produced.

The levies are calculated on both nitrogen and phosphate produced and the amounts are the same for each kg N and P_2O_5. The levy amounts are:

- For the quantities that fall within the first bracket: Bfr 1.25/kg
- For the quantities in the second bracket: Bfr 1.75/kg
- For the quantities in the third bracket: Bfr 2.25/kg
- For the quantities in the fourth bracket: Bfr 3.00/kg

Transfer levy

A transfer levy has to be paid on all manure that is transferred through the mediation of the manure bank. The levy is calculated on the basis of the cost of transport, storage, processing and administration. Depending on the quality of the manure, the levy can be lowered with a premium amount, for high quality manure, or increased, for low quality.

Import levy

Every importer of manure pays a levy of Bfr 100/ton.

A number of subsidies can also be applied for. All subsidies only apply for special protection areas, that is areas in the water group and in the nature group. No subsidies are available for phosphate-saturated areas. All subsidies only apply to agricultural land.

1. Subsidy for lower yields in special protection areas. Subsidies can be applied for in areas with lower fertilization limits. For grasslands in these areas, subsidies are different for areas with dairy cows and those without. There is also a limit

Table 4.16 Four brackets for basic levy

Brackets	Weighted phosphate production (kg)	Weighted nitrogen production (kg)
Bracket 1	1500–5000 + WP$_p$*	3000–10 000 + WP$_n$*
Bracket 2	5000 + WP$_p$* to 10 000 + WP$_p$*	10 000 + WP$_p$* to 20 000 + WP$_n$*
Bracket 3	10 000 + WP$_p$* to 15 000 + WP$_p$*	20 000 + WP$_p$* to 30 000 + WP$_n$*
Bracket 4	more than 15 000 + WP$_p$*	more than 30 000 + WP$_n$*

*WP$_p$ = Weighted production of processed or exported P
*WP$_n$ = Weighted production of processed or exported N

per farm of the area with dairy cows that can be subsidized. Subsidies are higher in the water group than in the nature group. Subsidies for maize are a flat rate per hectare.

2. Subsidy for manure storage. In special protection areas subsidies can be given for manure storage. Only holdings with less than 6 months' storage can apply.
3. Subsidy for purchasing mineral fertilizer. Holdings in areas of the water group can apply for subsidies for mineral fertilizer for maize and 'other crops'.
4. Subsidy for distribution costs.
5. Compensation of equity loss. Land owners in areas with lower fertilization limits can apply for a compensation for their loss of equity when they sell their land. A comparison is made between the value of the land before the Manure Decree became operable and after.
6. Family livestock farms with agricultural land in nature areas can make a request to the manure bank to buy their land, if they are willing to sell.

EFFECT OF MEASURES

Data have been collected by PARCOM to evaluate the effects of the above measures on the amount of nutrients being applied by agriculture. These show that average manure application rates in Belgium between 1985 and 1992 have increased in total-N (222–252 kg N/ha/year), and in total-P (50–53 kg P/ha/year) (see Table 3.13). Application rates of mineral fertilizer have been reduced by 8% for total-N and by 24% for total-P (see Table 3.14). The nitrogen surplus has increased 6% in this period, the phosphorus surplus has decreased 5.5% (see Table 3.15).

Provisional estimates for Belgium show that the expected reduction in the total amount of nutrients from agriculture between 1985 and 1995 would be >32% for total-P, and >11% for total-N. The reductions given between 1985 and 1992 are 32% for P, 11% for N (see Table 3.16). There is a considerable discrepancy between the data in Tables 3.13–3.15 and those in Table 3.16 that can only be explained by a change in land use: if the total area agricultural land decreased considerably between 1985 and 1992, that could explain the fact that application and surplus per ha changed little, while total inputs show a moderate to large decrease.

The evaluation of the Codes of Good Practice in relation to the Nitrate Directive that was commissioned by the EU shows the theoretical effects that measures A1–A5 of the Code could have on the quality of surface water in Belgium (Figure 4.5; de Cooman et al. 1995) if run-off measures are strictly adhered to. In the province of Antwerpen, the reduction in direct losses and run-off could have a major impact on surface water quality, in the order of more than 100 mg nitrate/l. In Oost- en West-Vlaandere and Limburg, the reduction would be in the medium range, around 40 mg/l. In Brabant and in Wallonia, the impact will be slight, somewhere around 20 mg/l.

The theoretical effects of measure A6 on surface water, shallow groundwater and deep groundwater is shown in Figures 4.7–4.9 (de Cooman et al., 1995). These figures show that the theoretical effect of measure A6, if implemented consistently,

70

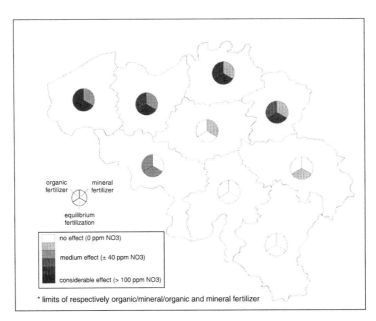

Figure 4.7 Theoretical effect of measure A6 on the quality of surface water. Source: de Cooman *et al.*, 1995

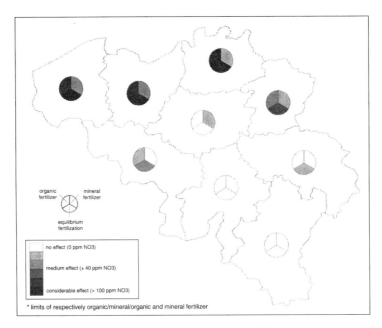

Figure 4.8 Theoretical effect of measure A6 on the quality of shallow groundwater. Source: de Cooman *et al.*, 1995

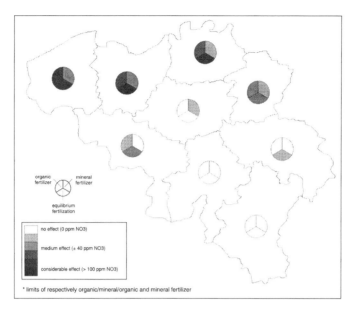

Figure 4.9 Theoretical effect of measure A6 on the quality of deep groundwater. Source: de Cooman *et al.*, 1995

will not have much effect on water quality in Wallonia. The theoretical effects in Flanders could be considerable, however: if the limits on organic, mineral, and organic and mineral both will be respected, a reduction in nitrates in surface water and in both shallow and deep groundwater could be 40–100 mg/l, and in some places more than 100 mg/l.

ROLE OF DIFFERENT ACTORS

During the negotiations on the Manure Action Plan, the Belgian Farmers Union (Belgische Boerenbond) has played a very important role in promoting the concept of the Family Livestock Farm, as a way to reorganize the structure of the Flemish agriculture, by imposing stricter rules on non-family livestock farms. Gaining some strategic advantage by negotiating positive discrimination for the family livestock farm, was also a way of unifying and consolidating a great number of farmers behind the Manure Action Plan and the manure reduction measures that ultimately had to be taken. The family livestock farms are producers who normally also cultivate some land. They therefore have less surplus and will use the land with more respect to the environment than livestock farms with no land, whose total manure production is a surplus.

This last argument also appealed to the environmental organizations, who included in their vision of the development of a sustainable agriculture in Flanders, based on a combination of ecological and social objectives. The Minister of the

Table 4.17 Indicative final fertilization limits in 2003 (kg per ha per year)

Main crop	Phosphate (P2O5)	Nitrogen (N)		
	All fertilizers (kg P2O5 total)	All fertilizers (kg N total)	Livestock and other organic fertilizer (kg N mineral)	Mineral fertilizers (kg N livestock)
Grassland	125	450	250	250
Maize	100	275	225	130
Crops with low N requirements	100	125	125	70
Other crops	100	275	200	150

Environment in Flanders invited the environmental organizations to take part in the political debate on the Manure Action Plan. Since this was the first time that the environmental organizations were asked to participate in a political process as equal partners they could not refuse the offer. In the end, through long negotiations, a win-win situation was found in combining environmental and social goals.

FUTURE DEVELOPMENTS

Since new comprehensive legislation is still recent it will be important to evaluate the impact of the new measures and to monitor environmental improvements when they occur. The impact of the present norms will be evaluated in 1998. The intention of the Manure Decree is to tighten the limits of both nitrogen and phosphates. If no compromise can be reached on lower limits, the limits will be lowered automatically every year from 1999 by 6 kg phosphate and 6 kg nitrogen, until the final limits have been reached. The proposed limits for the year 2003 are given in Table 4.17

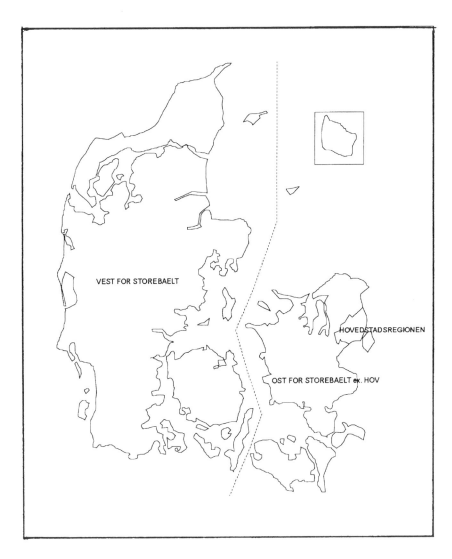

VEST FOR STOREBAELT

HOVEDSTADSREGIONEN

OST FOR STOREBAELT ek. HOV

DENMARK

CHAPTER 5: DENMARK

COUNTRY CONDITION

Denmark is a small country: it occupies about 4 million ha, of which about 2.8 million ha is utilized by agriculture. The agricultural sector involves around 74 900 farms and employs 160 000 people. Most holdings are family-type farms and fairly small: most farms are between 15 and 60 ha. However, between 1982 and 1991 farms have become more specialized. During this period, the number of holdings decreased from 98 500 to 74 900 and the average size has increased from 28.8 to 36.7 ha. Currently, almost 20% of the agricultural area is held by holdings of more than 50 ha. The value of agricultural production amounts to 6.851 million ECU (de Cooman, 1995).

The socioeconomic importance of agriculture is considerable: 65% of the total agricultural production is exported and agriculture accounts for 20% of all exports (goods and services) and employs about 10% of the total workforce. Between 1960 and 1988 agricultural production increased by 45%; crop production (sales) increased by 70%, while animal production increased by 36%, a rather modest level of growth compared to other European countries. In Denmark, livestock densities have not exceeded tolerable levels from a waste point of view (Baldock and Bennett, 1991).

About 93% of the total agricultural area consists of arable land, 60–70% of which is utilized for growing cereals. Permanent grassland covers only 3–8% of the total area. Denmark has around 4 million LU or 4% of the European livestock population. This population consists of 38% cattle, 56% pigs and 5% poultry. The stockbreeding sector is becoming increasingly specialized and farms are increasing in size: in 1991, 7% of the specialised holdings possessed 71% of the cattle and 75% of the pigs. Between 1982 and 1992 the cattle population decreased 22%, while the pig population increased by 14% and the average number of pigs per holding increased from 168 to 345. The importance of poultry has not changed a great deal (de Cooman *et al.*, 1995). Dairy production has moved from the eastern part of the country to the west (Western Jutland). The density of the pig population is spread more evenly across the whole country, but the greatest densities are also found on Jutland (Fig. 5.1).

ENVIRONMENTAL PRESSURES AND SOURCES

Use of mineral fertilizer

In Denmark, the average utilization of mineral fertilizer is about 140 kg/ha and is sufficiently high to rank somewhere in the middle regions in comparison with other

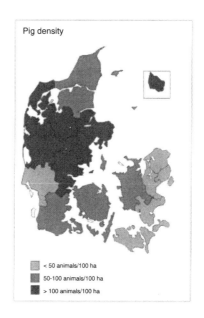

Figure 5.1   Animal density in Denmark

EU countries (Table 5.1). Table 5.2 shows the quantities that were used between 1975 and 1991. Even though the total consumption of mineral fertilizers has not changed much during that period, the use of nitrogenous fertilizer has increased from 1 218 000 to 1 457 000 t, or 357 000 to 395 000 t of N (de Cooman *et al.*, 1995).

Use of animal manure
The quantities of N produced by Danish livestock has been calculated per region (Table 5.3) and varies from 44.4 to 118.3 kg N/ha. Manure production is thus highest in Vest for Storebaelt. Table 5.4 shows shows the combined utilization of mineral and organic fertilizer in the different regions; again, the highest surplus is found in West Jutland. For Tables 5.1, 5.3 and 5.4 actual use is compared to the standard of 170 kg N/ha, recommended in the Nitrate Directive (De Cooman *et al.*, 1995).

Table 5.1   Mineral fertilizer (kg N/ha x year)

NUTS II	Limit	Utilization	Difference
Vest for Storebaelt	140.7	141.0	0.3
Copenhagen and surroundings	140.8	146.9	6.1
Ost for Storebaelt	125.6	141.0	15.4

Source: de Cooman et al., 1995

Table 5.2 Utilization of mineral fertilizer (1000 tons/yr)

Fertilizer	1975/76 –79/80	1980/81 –84/85	1988/89	1989/90	1990/91
Calcium nitrate, 15.5% N	47	25	24	21	21
Boron-calcium nitrate, 15.5% N	4	2	1	1	
Sodium nitrate				7	18
Sulphate of ammonia, 21.0% N	4	2	2	3	6
Calcium ammonia nitrate, 26.0% N	112	154	397	454	491
Liquid ammonia, 82.2% N	170	158	98	83	76
Urea, 46.0% N	6	15	11	20	16
Other fertilizers				4	6
NPK 16–5–12 (+Mg)	103	44	23	19	16
NPK 22–2–12 (+ Mg)	18	100	89	70	
NPK 21–4–10 (+ Mg)282	229	192	194	172	
NPK 25–2–6 (+ Mg + B)	312	548	233	199	165
NPK 25–3–9			180	210	199
NPK 23–3–7 (+ Mg + Cu)	90	47	26	27	24
Other NPK fertilizers	75	30	57	142	175
Garden fertilizers	13	5	2	2	2
Total	1218	1277	1346	1475	1457

Table 5.3 Organic fertilizer (kg N/ha x year)

NUTS II	Limit	Production	Difference
Copenhagen and surroundings	156.3	44.4	−111.9
Ost for Storbaelt, ex. Copenh.	154.1	58.4	−95.7
Vest for Storbaelt	150.7	118.3	−32.4

Source: De Cooman et al., 1995

Table 5.4 Organic and mineral fertilizer (kg N/ha x year)

NUTS II	Limit	Utilization	Difference
Copenhagen and surroundings	140.8	169.1	28.3
Ost for Storebaelt ex Copenh.	125.6	170.2	44.6
Vest for Storebaelt	140.7	200.2	59.4

Source: De Cooman et al., 1995

Average utilization of manure for the country as a whole is 102 kg N/ha, and is slightly higher for Jutland and Fünen (Vest for Storbaelt) at 116 kg N/ha (see Table 1.1). Fertilizer use is 141 kg N/ha, both for the country as a whole as for Jutland and Fünen (Schleef and Kleinhanss, 1996). Research at farm level shows that in Denmark

26% of the holdings supply animal manure at levels exceeding 170 kg/ha (see Table 3.3). Together these holdings are responsible for 59% of the total manure production; their average supply of N is 258 kg/ha (Brouwer *et al.*, 1995).

Green cover in winter

As Figure 5.2 shows, more than 70% of the total agricultural surface was covered with some type of winter crop in 1991/92 and this proportion is increasing: this is partly due to the fact that winter cereal crops are increasing. The area that is covered with 'green fertilizer', that is crops that are grown specifically because their uptake of nitrogen out of the soil, are grown on only 10% of the agricultural area. The Danish Code of good agricultural practice demands some type of winter crop on 65% of the area.

Nitrogen balance

The average nitrogen balance for the country as a whole is 104 kg N/ha. For the area with the most intensive livestock farms, Vest for Storebaelt, the average balance is 113 kg N/ha, which is not much higher than that for the whole country (see Table 1.3). Nitrogen balances for the country as a whole are moderate, but the averages may cover large differences between farms, with the most intensive livestock farms showing much higher nitrogen balances (Schleef and Kleinhanss, 1996). For example, average supply of N from manure for the most intensive livestock farms is 258 kg N/ha (see Table 1.1) compared with 102 kg N/ha for the country as a whole (see Table 3.3; Brouwer *et al.*, 1995).

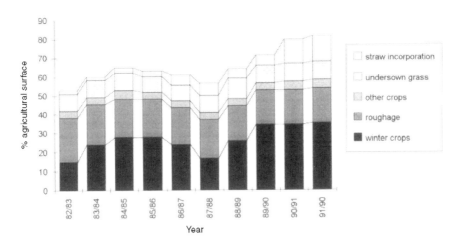

Figure 5.2 Green cover during the winter straw incorporation; undersown grass; other crops; roughage; winter crops

Phosphorus

Phosphorus from agricultural sources is not a problem in Denmark. Phosphorus pollution is attributed mainly to municipal waste water from urban areas.

AMBIENT ENVIRONMENTAL CONDITIONS OF SOILS, SURFACE-, GROUND- AND DRINKING WATER

The action plan that was introduced in 1987 in order to meet the standards of the EU Nitrate Directive also contains a program for intensive sampling of surface- and groundwater, in order to establish levels of pollution and monitor changes therein.

Surface water

Table 5.5 shows shows that there is a large difference in nitrogen concentration between water basins in natural environments, in agricultural areas and basins that are receiving point sources. It also shows differences in the amount of nitrogen in run-off into these different basins. The average concentration of N in agricultural areas is seven times that in natural environments. Quality objectives for surface water in relation to N were met at only 38% of the sampling stations. Data on changes in the concentration of N total in surface waters in Denmark, show no decreasing concentrations between 1989 and 1992. It should be noted that exceptional climatic conditions in 1992, notably a very dry summer and a very wet autumn, may have exaggerated the 1992 data (Table 5.6).

Data from another study show that between 1989 and 1992, the average levels of N in sources in agricultural areas are about ten times as much as those in natural areas and there are no significant changes in concentration over that period (Table 5.7).

Groundwater

About 1000 sampling stations have been selected in the framework of the Nationwide Monitoring Programme (NWMP), to monitor nutrients in groundwater in 68 regions. Results of this monitoring program show that two groups of regions emerge: Jutland and Northeast Zealand on the one hand, the rest of the country on the other. Figure 5.3 shows the concentration of nitrates per liter in aquifers

Table 5.5 Average concentration of N total

Type of basin	Average concentration of N total (mg/l)	Run-off (kg/ha)
Natural environment	1.0	2.1
Agricultural environment	6.9	18.7
Point sources	7.2	21.4

Table 5.6 Quality of surface waters

Year	Average concentration (mg N-total/l)
1989	5.8
1990	6.8
1991	6.3
1992	7.1

Table 5.7 Concentrations mg N-nitrate/l

Year	Basins in natural areas	Basins in agricultural areas
1989	0.55	5.36
1990	0.56	5.76
1991	0.63	5.90
1992	0.60	5.58

Sources in different areas

that have been sampled. Higher concentrations are found on Jutland and NE Zealand than in other parts of the country: up to 50 mg NO_3/l and in quite a few places over 50 mg/l.

Drinking water supply in Denmark is almost totally (98%) based on groundwater. Most drinking water is still of good quality, the nitrate content is well below the EU standard of 50 mg/l. Drinking water quality was improved somewhat in the 1980s by giving up aquifers with excessive nitrate content and drilling new and deeper boreholes to aquifers that are less affected by nitrate pollution.

Lakes and marine waters
The average N-total in lakes has increased from 2.70 mg/l in 1989 to 3.18 mg/l in 1992, which means that the water quality in lakes has decreased in that period. The water quality of marine waters has been monitored by the National Agency of Environmental Protection and the National Environmental Research Institute. Research has shown that the concentration of nitrogen during the winter is the limiting factor for the development of algae during the following summer. From the 1970s the level of nitrogen in marine waters increased steadily and stabilized during the late 1980s. Figure 5.4 shows the levels of N in marine waters in 1989. Figure 3.4 shows which zones are susceptible to eutrophication.

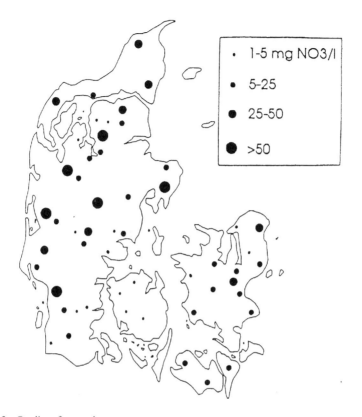

Figure 5.3 Quality of groundwater

NEED FOR AND PURPOSE OF EXISTING POLICY AND REGULATION (INTERNATIONAL AND NATIONAL)

Until the early 1980s there was little public concern about nutrient pollution of ground- and surface water. Early legislation (before 1974) concerned surface run-off and emissions from drainage systems from silage, farmyard manure and slurries. Pollution from these point sources resulted occasionally in considerable fish deaths in rivers and fish farms. In 1981 regional emergency plans were required in order to trace polluters more quickly, gather evidence and thus have a better change of convicting offenders. Nitrate and phosphate pollution of the aquatic environment by non-point sources was not really perceived as a problem. The main reason for this lack of concern is probably the fact that drinking water in Denmark is extracted from geologically well-protected groundwater sources, with relatively low levels of nitrates. This means there was no immediate health hazard to stimulate a public debate (Baldock and Bennett, 1991).

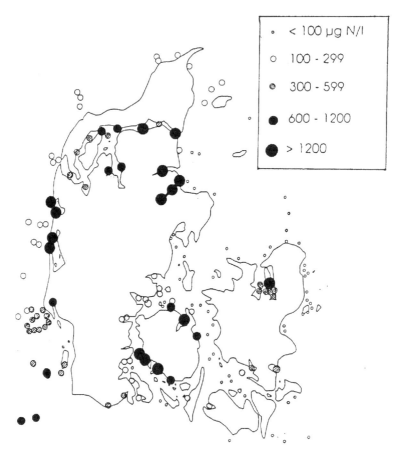

Figure 5.4   Concentration of nitrogen in sea water

International policy
International concerns about potential nutrient pollution problems and
the possibilities of eutrophication and groundwater pollution, in the context
of the INSC, where Denmark was one of the participants, has brought the
issue to the public's attention, especially the quality of the marine environment.

Denmark has accepted all rules and measures relating to manure policy of the
INSC conferences. It has also accepted all of the 1988 and 1989 PARCOM recommendations and four of the six 1992 measures. The status of the Nitrate Directive is that transposition has been notified, the whole territory has been designated
as vulnerable zone, the Code of good agricultural practice and the Action Program
have been notified.

National policy

Marine waters are very important to Denmark, both economically (fisheries and tourist industry) and for recreation. Deterioration of the marine environment can also be very tangible (fish deaths), with the result that quality concerns about surface water strike a responsive chord, both politically and in the public opinion. This has resulted in very far-reaching and ambitious policy goals and probably the most comprehensive abatement programme in Europe. Between 1985 and 1991, three major Action Programmes were accepted and implemented, with very ambitious policy goals: reduction of nitrogen losses of 50% and phosphate losses of 80% within 3 years.

A nationwide monitoring programme was also set up, to provide necessary data on the progress (or lack of it) being made. Thus, the Danish government showed that it was serious in its intentions to reduce nutrient pollution and to keep track of the results of its efforts.

POLICY FORMULATION AND LEGAL INCORPORATION

In 1985, Parliament accepted the so-called NPO Action Plan to abate pollution of the aquatic environment by nitrogen, phosphorus and organic substances. In 1987, this plan was replaced by the more comprehensive Action Plan on the Aquatic Environment. The policy goals of this plan were to reduce nitrogen losses with 50% and phosphate losses with 80% within 3 years. The main objectives of this plan were to stop run-off and leaching of manure and silage, to ensure adequate storage capacity for manure and to establish the 'harmonization rule', which matches herd size with the land available: a sufficient area must be available for the spreading of farmyard manure, in order to prevent very high livestock densities on farms.

In 1988, the Nationside Monitoring Programme (NWMP) was set up to demonstrate the effects of policies and investments that have been the result of the Action Plan on the Aquatic Environment. By systematic sampling, it should be possible to get accurate information on the sources of nutrients that pollute ground- and surface water. It also measures atmospheric depositions. Its main objective is to assess the damage that is done to ecosystems by nutrient emissions. When it became apparent that the policy targets that had been formulated as part of the Action Plan on the Aquatic Environment were not going to be met, a new plan was formulated, the Programme for a Sustainable Agriculture. This plan was introduced in 1991 and from spring 1993 the rules were enforced. The programme aims at a better utilization of manure and ties the total amount of nitrogen applied to the estimated need for nitrogen at the farm. This requires fertilizer and crop rotation plans from farmers to realize this goal. Additional measures were agreed upon later, to meet EU Nitrate Directive standards.

Another policy goal of this plan is to anticipate a situation where the agricultural sector will receive fewer government subsidies or will completely go without and to stimulate developments that will facilitate its survival under changed circumstances.

REGULATORY CONTROLS

The Action Plan on the Aquatic Environment
This plan is very comprehensive in scope, as it deals not only with agriculture, but also with water treatment plants, industrial emissions, fish farms, power plants, motorized vehicles and pollution of Danish waters by non-Danish sources. Measures concerning agricultural emissions are the following:

- Harmonization rule: application of animal manure and wastes to land should not exceed the capacity of crops to utilize the nutrients. Therefore, upper limits for the application of manure have been set. These are the equivalent of 2.3 LU/ha on dairy farms and 1.7 LU/ha on pig farms. Farms with higher livestock densities may prove compliance with the standards by presenting written agreements with other farms to receive their surplus manure.
- Application of manure is generally prohibited from harvest until 1 February.
- A minimum of 9 month storage capacity for animal manure is required for farms with a minimum of 31 LUs. In 1988, the requirement for 9 months' capacity has been adjusted to 6–9 months, depending on individual circumstances. For smaller farms, there are no specific rules, but a general provision demanding sufficient storage capacity to ensure environmentally sound manure application.
- A minimum of 65% winter crops on all farms.

Action Plan for a Sustainable Agriculture
This plan aims at a better utilization of manure by setting minimum utilization standards of farmyard manure. These standards were then combined with rules that prescribe that the total amount of nitrogen applied at farm level (in fertilizer and manure) must not exceed the estimated need for nitrogen by the crops. Thus, farmers must prepare yearly crop rotation and fertilizers plans, comprising the following:

- A crop plan, including green cover areas.
- A fertilizer plan, stating the needs of the crops for nutrients and how the needs are met, using farmyard manure and/or commercial fertilizer. When estimating need, standards fixed by the Danish Plant Directorate can be applied. These standards can be adjusted to individual circumstances.
- The fertilizer balance must state information on the need for nitrogen and the consumption of N from farmyard manure and other fertilizers.
- The total nitrogen application must not exceed the amount needed, such that the following minimum utilization percentages can be met for nitrogen farmyard manure:

Pig slurry, until 1–8–97/after 1–8–97	Min. 45/50%
Cattle slurry, until 1–8–97/after 1–8–97	Min. 40/45%
Manure from deep litter systems, from 1–8–93	Min. 15%
Other types of farmyard manure, from 1–8–95	Min. 40%

Other measures include:

- Methods and periods for the application of manure:
  Liquid farmyard manure may not be applied between harvest and 1 March, with some exceptions.
  Liquid farmyard manure and silage effluent spread on areas with no vegetation are to be incorporated as soon as possible and within 12 hours.
  Solid manure can only be spread in the period between harvest to 20 October on areas covered by crops the following winter.
  Solid manure spread on areas with no vegetation is to be ploughed in immediately after spreading.
- Measures concerning storage capacity for manure and winter crops are maintained.
- Other measures have to do with the protection of groundwater in sensitive areas, where groundwater is used for drinking water and the nitrate concentration is more than 50 mg/l. Compensation will be given to local farmers for reduced fertilizer use.
- Freshwater wetlands can act as filters for N through the denitrification process; pilot projects will be started to investigate the possibility of revitalizing wetlands for this purpose.
- The EU will reimburse member states that are willing to let arable land lie fallow, in order to decrease wheat production. This may be implemented for 10–15% of sandy soil areas, that have a greater risk of polluting groundwater with nitrogen.

Environmental Licensing

All pig and poultry farms with at least 250 LUs must apply for a 'VVM evaluation' (environmental impact evaluation) and a special environmental approval, in case of the establishment or expansion of such farms. Standards have to be met for siting and construction of livestock housing, dung pits and other silage and manure storage facilities. Applications have to be sent to the municipality, who is responsible for the licensing process and control.

Code of Good Agricultural Practice

By far the majority of the measures that have been proposed by the Nitrate Directive and are to be part of the Code of Good Agricultural Practice were implemented in Denmark some years ago (Table 5.8).

Table 5.8 Code of Good Agricultural Practice: Denmark

Code	Description	Stimulated by
A1. Periods when application of fertilizer is inappropriate	Solid manure: 20/10 – 01/02 Liquid manure: harvest – 01/01 Silage effluent: <01/11	Legislation and regulation
A2. The land application of fertilizer of steeply sloping ground.	No direct emissions, obligation to plough down immediately.	Legislation and regulation
A3. The land application of fertilizer to water-saturated, flooded, frozen or snow-covered ground.	Inundated soil: All conditions, see Frozen soil: also A1; immediate Snow-covered soil.: plough down	Legislation and regulation
A4. The conditions for land application of fertilizer near water courses.	No cultivation within 2 m. No direct emissions	Advice, Legislation and regulation
A5. The capacity and construction of vessels for livestock manures, including measures to prevent run off and seepage.	At least 6 months, usually 9 months required; possibility for exceptions.	Legislation and regulation, Subsidies
A6. Procedures for the land application of fertilizers to maintain nutrients losses to water at an acceptable level.	Economic optimum, lower in protected areas.	Advice, Legislation and regulation
B7. Land use management	Compulsory crop rotation plans; Max. 2.3 livestock units/ha.	Legislation and regulation
B8. Use of catch crops	65% autumn crop cover required.	Legislation and regulation
B9. Establishment of fertilizer plans and keeping records of fertilizer use.	Nitrogen balance per farm	Compulsory fertilizer balance sheets, levies

MONITORING AND CONTROL, RECORD KEEPING AND VERIFICATION

Monitoring of ground- and surface water

The quality of ground- and surface water is monitored by the Nationwide Monitoring Programme (NWMP), by systematically gathering data on water quality from lakes, rivers, coastal waters and ground water. Information on pollution sources and their effect on the aquatic environment provides feedback on the effectiveness of the measures that have been implemented so far.

Implementation and enforcement of Nitrate Policy Measures

In June 1994, it was decided that all farmers had to supply evidence of sufficient storage capacity for livestock wastes before the end of the year. As Fig. 5.5 shows, the percentage of farmers with 9 months' storage capacity had increased from less than 40% in 1988 to 69% in 1994. Also, 93% of all farmers had at least between

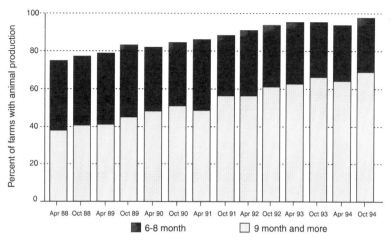

Figure 5.5 Storage capacity for animal manure. Source: Danish Farmers' Association

6 and 8 months' capacity, so compliance with this measure is very high and is facilitated by financial measures.

Compliance with the Harmonization Rule, which stipulates how to match herd size with land available, is shown in Fig. 5.6. It shows that most disharmonic farms are in the pig production sector, but an average of more than 80% of all farms comply with the rule. However, the numbers shown in Fig. 5.6 are somewhat misleading, since the rule is concerned with the amount of manure produced on the farm, not the amount applied. Disharmonic farmers can make contracts with other farmers without harmony problems to take their manure surplus and these

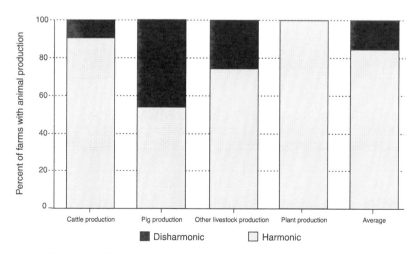

Figure 5.6 Distribution of harmonic and disharmonic farms, 1994. Source: Danish Statistical Bureau

transfers are not reflected in the data. Compliance is, therefore, underestimated in this figure (Schou, 1997).

Fertilizer plans with a fertilizer balance have to be filed every year before 1 August. In 1993/94 about 30 000 spot-checks were made to check on their accuracy: these checks were partly randomized, but more than half concerned farms with large livestock populations. Farmers who did not achieve a sufficiently high utilization of their manure were fined. Table 5.9 shows shows the percentage of farms that deviate from the minimum utilization standard in either direction (Schou, 1997).

On the basis of these spot-checks, it could be concluded that there is broad compliance with the rules concerning the utilization of manure, with an average utilization of 45%. However, it also became clear that some of the key figures used can be legally adjusted by the farmer: the figure for nitrogen need is based on the farmers' expectations of the yield of the coming harvest and can be adjusted upwards if justified and documented by a consultant in plant production. Farmers need not keep yield records at all. Cross-checks of expected yields with regional data of yields show that yields on the average have been overestimated by about 30%, nitrogen need by 10% and utilization of manure by 45% (based on calculations by Schou, 1997). According to Schou, farmers should have less leeway in adjusting their expected yield figures, so they seem to comply on paper and thus avoid fines. Schou proposes that standards should be either set on a regional basis or the expected yield should be equal to the average yield of the last 3 years (Schou, 1997; Frederiksen and Schou, 1996).

The agricultural organizations in Denmark were initially caught by surprise when, in the mid-1980s, environmental issues in agriculture became a political issue. Initially, they denied all connections between agriculture and environmental problems, but that attitude has changed. Since then, they have participated constructively in the political process of implementing environmental regulations for agriculture. They have also become much more powerful in influencing environmental policies. For example, they were able to modify initial requirements for storage capacity for manure. Instead of the 9 months' capacity requirement, they reached a compromise that requires individual assessment of the need

Table 5.9 Distribution of farms after their deviation from the minimum utilization of manure

	Deviation from minimum utilization of manure						
Less than −20	−20 to −10	−10 to −5	−5 to 0	0 to 5	5 to 10	10 to 20	>20
Percentage of farms	5.0	4.8	6.5	25.5	13.6	14.7	25.3

Source: The ministry of Agriculture and Fisheries; The directorate for Crop Production.

On basis of the 1993/94 Key-figures only a small number of farmers who had violated the minimum standards heavily were given a fine. Their lawsuits are currently being tried in court to set a precedent for future cases.

for storage capacity for 30 000 or so livestock farms. It soon became clear that there was no political will to furnish environmental control agencies with the resources to enforce this legislation. Since then, the responsibility for implementing storage capacity has also been transferred to the Agricultural Advisory Service, which is run by the agricultural organizations. Enforcement of measures is, therefore, partly in the hands of the agricultural sector itself (Baldock and Bennett, 1991).

One measure that can be effectively controlled is leaching and run-off from storage of animal waste. Officials from municipalities make control visits to farms and will do repeated inspections, if necessary. All facilities need to be inspected once every 10 years. Non-compliance will be prosecuted and fined.

FINANCIAL MEASURES AND INCENTIVES

Policy measures in Denmark have used financial instruments such as levies and fines rather cautiously: fines are given for insufficient utilization of manure, but only to the worst offenders. Politicians tend to avoid negative effects on farmers' incomes. Interesting in this context is the nitrogen tax debate: in 1987, the environmentalists in Parliament passed a resolution demanding that a tax of about 8% should be imposed on nitrogen in commercial fertilizer, in order to finance subsidies for the improvement of animal wastes. This caused a lot of protest from the side of agriculture and the proposal was withdrawn. Later, a tax on nitrogen was again proposed, in case agriculture failed to meet the required reduction in nitrogen use. When in 1989 it became clear that nitrogen consumption had not decreased, the Social Democratic party demanded immediate implementation of the tax, but this proposal did not gain sufficient parliamentary support. Instead, new investigations were decided on, to research the measures that would be needed to ensure a 'sustainable agriculture' (Baldock and Bennett, 1991). As was mentioned above, farmers can and will be fined for non-compliance with waste storage facilities, which are controlled by the municipality. Construction, expansion and improvement of manure storage facilities are subsidized, with 35% of the construction costs being financed by public grants and loan guarantees being given by the Danish State Mortgage Bank. Investments in storage facilities amounted to about 10% of total investments in agriculture between 1986 and 1994. In groundwater protection areas, farmers are compensated for the losses they incur, by applying less fertilizer than the economic optimum. Other financial measures are subsidies to organic farming and payments under the EU Directive 2078 (accompanying measures).

EFFECT OF MEASURES

As has been mentioned above, the measures taken have generally been well implemented by farmers, even though the implementation of measures is not always as strict as it should be. There is a general tendency among farmers to overestimate the nutrient needs of the crops for the coming year, resulting in applications of more fertilizer than is really necessary. On the other hand, storage capacity for manure has generally been implemented and leaching from silos and storage facilities has really diminished.

In 1990, the Danish Ministry of the Environment estimated the environmental effects of the nitrogen reduction measures for agriculture and came to the conclusion that nitrogen emissions from agriculture had been reduced by 20%, since the beginning of the Action Plan for the Aquatic Environment. Estimated reductions are shown in Table 5.10.

The total estimated emissions of nitrogen from agriculture were about 250 000 tons N/year before the beginning of the Action Plan, so a reduction of about 20% was realized between 1987 and 1990; this reduction was considerably below the expected 50%, which was the target of the Action Plan. It was also estimated that the ploughing in of animal manure has reduced the ammonia volatilization by about 20 000 tons/year. Ammonia volatilization was, before the Action Plan for the Aquatic Environment, about 95 000 tons N/year, so here too a reduction of over 20% has been realized (Ministry of the Environment, Denmark, 1991).

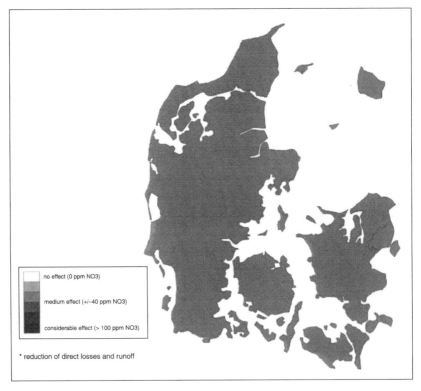

Figure 5.7 Theoretical effect of measures A1–A5 on the quality of surface water. Source: de Cooman *et al.*, 1995

Table 5.10   Estimated reductions in nitrogen, 1987–1990

Total reduction	(tons N/year)
Crop rotation, catch crops, green fallowing	15 500
Reduced leaching through straw ploughing	3 000
Planning, handling, utilization of manure	11 500
Decrease in arable land	5 000
Leaching from silage and storage facilities	15 000
Total	50 000

Data from PARCOM show that the application rates of manure for the period 1985–1992 fell from 100 to 91 kg N/ha/year, and from 20 to 17 kg P/ha/year (see Table 3.13). For mineral fertilizer, these figures show a 6% reduction for total N, a 29% reduction for total-P (see Table 3.14). The nitrogen surplus showed no reduction for total-N and total-P between 1985 and 1990 (see Table 3.15). The reduction of all inputs of nutrients from agriculture did not change between 1985

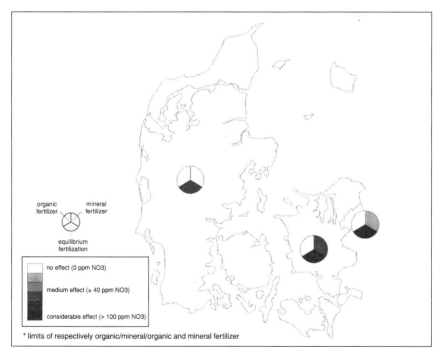

Figure 5.8   Theoretical effect of measure A6 on the quality of surface water. Source: de Cooman *et al.*, 1995

and 1992, but a 5% reduction was expected for total-N between 1985 and 1995, with a 15% reduction for total-P in that same period (Table 3.16).

Research on water quality has shown that both the quality of surface waters and of groundwater have stabilized: water quality has not yet improved, but it has not deteriorated either. Noticeable changes for the better will take time, as experts agree; also, the variability of climatic conditions can have major effects on run-off and leaching and this makes the results of the National Monitoring Program difficult to interpret.

The evaluation of the Codes of Good Practice in relation to the Nitrate Directive, commissioned by the EU, shows the theoretical effect that Measures A1–A5 could have on the quality of surface water in Denmark (Fig. 5.7). The effects will be in the medium range (± 40 mg/l NO_3) for the whole country (De Cooman *et al.*, 1995). The theoretical effect of measure A6 on surface water, shallow groundwater and deep groundwater is shown in Figures 5.8–5.10, where the effects of equilibrium fertilization are expected to be strong (> 100 mg/l NO_3) for all conditions and medium effects can be expected from reduced use of mineral fertilizer (de Cooman *et al.*, 1995).

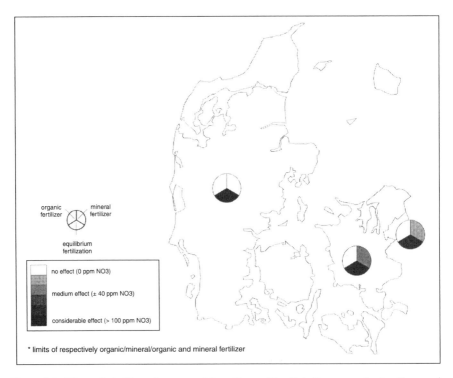

Figure 5.9 Theoretical effect of measure A6 on the quality of shallow groundwater. Source: de Cooman *et al.*, 1995

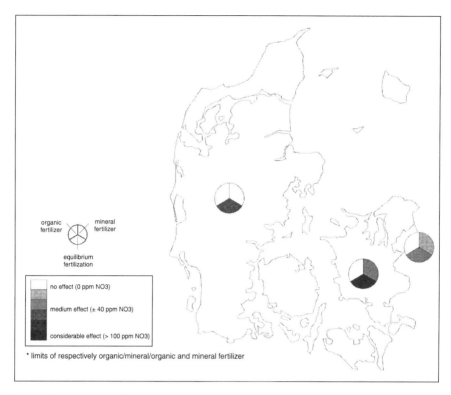

Figure 5.10   Theoretical effect of measure A6 on the quality of deep groundwater. Source: de Cooman *et al.*, 1995

ROLE OF DIFFERENT ACTORS

In the early 1980s, the environmental organizations in Denmark were very strong: they were well represented in Parliament and had a lot of political clout. The acceptance of the NPO Plan in 1985 and the more comprehensive Action Plan for the Aquatic Environment in 1987, which set strict standards for agriculture to reduce nutrient pollution, were very much due to their influence. Since then, the agricultural organizations have gained a much stronger influence on environmental politics and have much more control over the implementation and administration of these policies. For example, the assessment of storage capacities on individual farms is now under control of the Agricultural Advisory Service. The refusal to implement the proposed nitrogen tax has been a victory for the agricultural lobby. Furthermore, the drafting of the Programme for a Sustainable Agriculture, initially the responsibility of the Minister of the Environment, was then transferred to the Minister of Agriculture: where political power originally favoured the environment, it has now shifted towards agriculture (Baldock and Bennett, 1991).

FUTURE DEVELOPMENTS

Even though Denmark already has an ambitious programme for the reduction of nutrients in ground- and surface water, it has also become clear that the objectives of a reduction of 50% will not be met by present legislation. One additional measure that will be implemented from 1 January 1997 is that the area requirements for livestock farms will increase: these new rules concerning the harmony between livestock hold and total agricultural area, prescribe that all farms with animal production should own a minimum percentage of the land, which is required to dispose of their manure. This minimum percentage will increase with the size of the livestock hold. The rule will apply to farms traded after 1 January 1997 and has already led to an increase in the demand for land. Selling farms and taking out mortgages will depend on meeting these area requirements in the future. With this new measure, Denmark has made clear that it wants to prevent the development of very intensive livestock holdings with insufficient areas to apply the quantities of manure that are produced on the farm. In fact, it is tightening the 'rule of harmony' by requiring a relatively greater ownership of land than before.

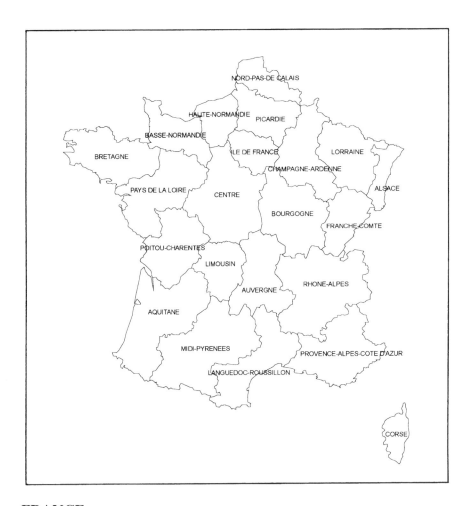

FRANCE

CHAPTER 6: FRANCE

COUNTRY CONDITION

France is the biggest country in Western Europe. It occupies about 55 million ha, of which 31 million ha (56%) is devoted to agriculture. There are about 982 000 farms, employing about 1 400 000 people. The value of its total production amounts to 46 000 million ECU and is thus the most important agricultural sector in Europe.

French agriculture is very diverse with considerable differences among the different regions. Some of the most important regions are:

- The western part of the country is dominated by livestock farms, especially Bretagne. Farms are mostly small, no more than 20 or 30 ha and have become more and more intensified over the last decade. In Bretagne, Haute-Normandie and the Loire region, fodder crops are important and occupy about 50% of the arable land.
- The area around Paris has large farms, often more than 100 ha, where cereals, potatoes and sugar beets are grown. South of this region and towards the east are large mixed livestock and dairy farms. Fodder crops and grasslands occupy about 50% of the area. It includes the Franche-Comté, Limousin and Auvergne.
- In the regions of Haute-Normandie, Ile-de-France, Lorraine, Nord-Pas-de-Calais, Bourgogne, Picardie, Champagne-Ardenne and Centre, about 50% of the arable land is used for cereals.

There are many different areas where vineyards, olive groves, fruit and vegetable farms, or maize are found. In short, French agriculture is very varied and very complex and there are great differences between the different regions in climate, soil and socioeconomic circumstances (de Cooman *et al.*, 1995).

The total livestock population in France is about 24 million LU, or around 24% of the European livestock population. In fact, it has more livestock than any other European country. The total livestock population consists of 70% cattle, 12% pigs, 10% poultry and 6% sheep and goats. Between 1983 and 1989 the livestock population decreased by 1.5 million LU, mostly cattle.

Bretagne has the largest livestock population (4.8 million LU), next is the Pays-le-Loire (3.1 million LU). The regions of Basse-Normandie, the Bourgogne, the Midi-Pyrenées, the Rhone-Alpes and the Auvergne all have more than 1 million LU. Only Bretagne has a density that is more than 2 LU/ha. Other areas have densities that are more than 1.5 LU/ha: Picardie, Haute-Normandie, Nord-Pas-De-Calais

98

and Alsace. Livestock farms that are not soil-bound are to be found in these regions (de Cooman *et al.*, 1995).

ENVIRONMENTAL PRESSURES AND SOURCES

Use of mineral fertilizer

Use of mineral fertilizer in France increased in the years 1970–1985 (Fig. 6.1) from a high of 120–160 kg N/ha in 1970 to a high of 160–200 kg N/ha in 1985. These maps also show that the areas where mineral fertilizer use is high has stayed roughly the same, but the general level has gone up. Table 6.1 shows more recent figures (1990) related to utilization of mineral fertilizer. According to these figures, mineral fertilizer use has decreased since 1985, but since these data stem from different sources, caution is required before drawing firm conclusions. Picardie is the area where most mineral fertilizer is used. Table 6.1 also shows that use of mineral fertilizer exceeds the limit in none of the regions (de Cooman *et al.*, 1995).

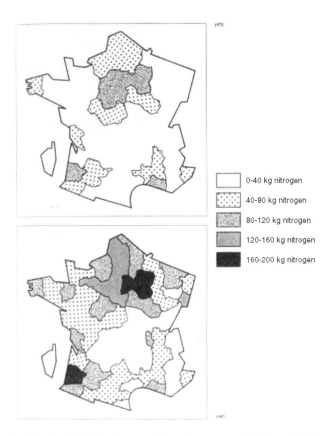

Figure 6.1   Analysis of artificial fertilizers used per Departement in 1970 and 1985

Table 6.1 Mineral fertilizer (kg/ha.yr)

NUTS II	Limit	Utilization	Difference
Corsica	188.3	5.0	–183.2
Limousin	186.6	18.0	–168.6
Auvergne	185.0	26.3	–158.7
Franche-Comte	182.4	42.5	–139.9
Rhone-Alpes	169.5	43.9	–125.6
Basse-Normandie	176.4	61.0	–115.4
Midi-Pyrenees	166.4	57.7	–108.7
Provence-Alpes-Côte d'Azur	134.4	31.7	–102.6
Bourgogne	162.2	64.1	–98.1
Languedoc-Roussillon	127.4	40.8	–86.6
Pays de la Loire	164.3	82.4	–82.0
Lorraine	173.8	93.5	–80.3
Aquitaine	162.5	83.0	–79.5
Alsace	168.2	113.9	–54.3
Bretagne	157.7	105.4	–52.4
Haute-Normandie	154.2	106.1	–48.0
Nord-Pas-De-Calais	165.7	120.4	–45.3
Poitou-Charentes	150.6	106.7	–43.9
Centre	138.7	108.1	–30.6
Champagne-Ardenne	151.2	123.9	–27.3
lle-de-France	136.1	129.2	–6.9
Picardie	144.0	144.7	0.7

Source: de Cooman *et al.* 1995

As shown in Table 1.2, the average use of mineral fertilizer for the whole country is 91 kg N/ha, whereas for Bretagne it is 108 kg N/ha (Schleef and Kleinhanss, 1996).

Use of animal manure

The production of animal manure has also increased between 1970 and 1985 (Fig. 6.2), but only for two regions, Bretagne (120–160 kg N/ha) and Basse-Normandie (80–120 kg N/ha). As Table 6.2 shows, manure production remained at that level in 1992 for these two regions and for Pays de la Loire. Manure production does not exceed the limit of 170 kg N/ha from the Nitrate Directive anywhere, even though it comes close in Bretagne at 150.5 kg N/ha.

Table 6.3 shows that, when the utilization of both animal manure and mineral fertilizer together is compared with the limit, Bretagne shows a considerable surplus, while regions like Picardie, Nord-Pas-De-Calais, Haute Normandie and Pays de Loire show moderate to low surplus levels. As shown in Table 1.1, average manure production for France is 52 kg N/ha, but in Bretagne it is 147 kg N/ha. Compared with countries such as the Netherlands and Flanders, this is a low to

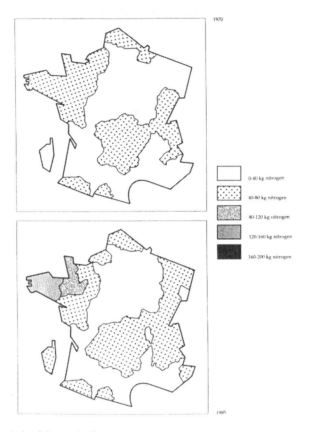

Figure 6.2   Analysis of livestock effluents used per Departement in 1970 and 1985

moderate level. However, when the analysis is performed at farm level and the farms with the highest level of manure production are singled out, we see that on the 6% of the holdings in France which produce animal manure at a level exceeding 170 kg N/ha, the average manure production is 309 kg N/ha (see Table 3.3).

Nitrogen balance

The average nitrogen balance for the whole of France is 63 kg N/ha, while for Bretagne it is considerably higher, at 130 kg N/ha. All the other regions in France have surpluses below 100 kg N/ha (see Table 1.3). In Bretagne and to a lesser extent in Pays de Loire, relatively high nitrogen balances are related to intensive livestock farming. In other more moderate areas, like Nord-Pas-De-Calais and Haute-Normandie, it is a combination of livestock farming and use of mineral fertilizer. Figure 6.3 shows the distribution of the different livestock populations in France: the pig, poultry and laying hen populations are all concentrated in the

Table 6.2   Organic fertilizer (kg N/ha.yr)

NUTS II	Limit	Production	Difference
Ile-de-France	168.9	7.5	−161.4
Corsica	165.9	13.7	−152.2
Languedoc-Roussillon	164.8	14.6	−150.2
Centre	169.3	20.0	−149.4
Provence-Alpes-Côte d'Azur	162.1	13.9	−148.2
Champagne-Ardenne	169.9	25.4	−144.5
Picardie	169.7	31.0	−136.6
Bourgogne	169.8	37.9	−131.8
Poitou-Charentes	169.7	38.8	−130.9
Midi-Pyrenees	168.6	38.8	−129.8
Aquitaine	167.3	42.4	−124.9
Alsace	169.2	45.1	−124.1
Lorraine	169.6	50.5	−119.1
Rhone-Alpes	165.7	47.6	−118.1
Franche-Comte	169.9	55.3	−114.6
Auvergne	169.8	57.3	−112.5
Haute-Normandie	169.6	58.3	−111.3
Nord-Pas-De-Calais	169.8	62.0	−107.8
Limousin	169.3	64.8	−104.5
Basse-Normandie	169.7	83.3	−86.4
Pays de la Loire	169.0	85.3	−83.8
Bretagne	169.6	150.5	−19.1

Source: de Cooman *et al.* 1995

Bretagne and the Haute- and Basse-Normandie regions, even though other parts of France show moderately high populations.

AMBIENT ENVIRONMENTAL CONDITIONS OF SOILS, SURFACE-, GROUND- AND DRINKING WATER

Surface water

The quality of surface water in France has steadily declined, as regular monitoring has shown. Table 6.4 shows that between 1971 and 1986, the percentage of measuring stations that have measured nitrate levels below 10 mg/l declined from 77.2 in 1971 to 48.4% in 1986. Over the same period the percentage of stations finding nitrate levels over 50 mg/l increased from 0.2% to 0.5%. This steady decline in rivers and river basins is unevenly distributed throughout the country. The worst affected rivers are those in Artois Picardie, the Seine Normandie basin and the western parts of the Loire Bretagne basin. More localized problems have been found in the Adour Garonne and the Rhine Meuse basins. Figure 6.4 shows the distribution of different levels of nitrates in surface waters in France, with most

Table 6.3 Organic and mineral fertilizer

NUTS II	Limit	Utilization	Difference
Corsica	188.3	18.7	−169.5
Limousin	186.6	82.8	−103.8
Auvergne	185.0	83.6	−101.4
Provence-Alpes-Côte d'Azur	134.4	45.6	−88.8
Franche-Comte	182.4	97.8	−84.6
Rhone-Alpes	169.5	91.6	−78.0
Languedoc-Roussillon	127.4	55.4	−72.0
Midi-Pyrenees	166.4	96.5	−69.9
Bourgogne	162.2	102.1	−60.2
Aquitaine	162.5	125.5	−37.1
Basse-Normandie	176.4	144.3	−32.1
Lorraine	173.8	144.0	−29.8
Centre	138.7	128.0	−10.6
Alsace	168.2	159.0	−9.2
Poitou-Charentes	150.6	145.5	−5.1
Champagne-Ardenne	151.2	149.3	−2.0
Ile-de-France	136.1	136.7	0.6
Pays de la Loire	164.3	167.6	3.3
Haute-Normandie	154.2	164.4	10.3
Nord-Pas-De-Calais	165.7	182.4	16.7
Picardie	144.0	175.8	31.8
Bretagne	157.7	255.9	98.2

Source: de Cooman *et al.* 1995

of the higher levels (> 20 mg/l) in the Bretagne, Loire and Picardie regions. In the coastal waters of Bretagne-Côtes d'Armor there have been serious eutrophication outbreaks, which interfere with tourism and create a general nuisance. The local population copes with this problem by physically taking algae out of the water, but even so disposal can be a problem.

A comparison of Figures 6.3 and 6.4 shows that high levels of nitrates in surface water roughly correspond with high levels of intensive livestock farming, and also with intensive crop farming (in Picardie, for example). Even though the intensity of livestock farming, even in Bretagne, is not as high as in the Netherlands and Flanders, it has caused serious pollution problems because farms and storage facilities are often old and inadequate and sometimes in disrepair. Since about 40% of the drinking water supplies come from surface waters, these high nitrate levels have consequences not only for eutrophication, but also for drinking water, which is even more serious from a public health point of view. Figure 6.5 shows nitrate levels in rivers in Bretagne, with the largest circles indicating higher concentrations, up to 88 mg/l in some places near the coast.

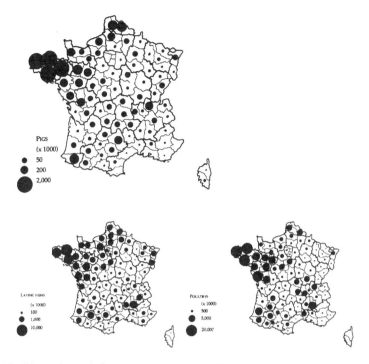

Figure 6.3 Livestock populations per Departement, 1988

Groundwater

Nitrate levels in groundwater and underground aquifers depend a lot on the soil type which the water has to pass through before it reaches the aquifer. In Bretagne, small, localized, unconfined aquifers can be found in areas of granite and schist (leisteen). These aquifers are heavily polluted by nitrates from livestock and crops, but they could lose their high nitrate levels rapidly (within a few years) because they are replenished quickly.

In the Paris basin and in Aquitaine, there are large sedimentary areas underground: these are more productive, but they are replenished very slowly. Nitrates may remain for a long time in the strata above the water table and leach

Table 6.4 Classification of measuring stations by nitrate concentration (%) in surface water

	<10 mg/l	*>50 mg/l*
1971	77.2	0.2
1976	67.2	0.3
1986	48.4	0.5

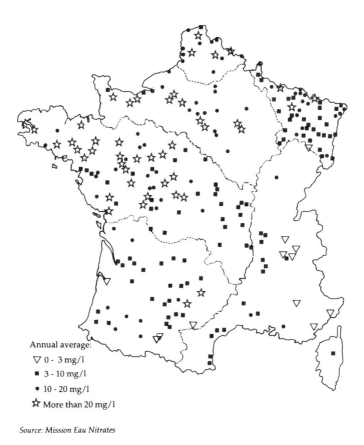

Annual average:

▽ 0 - 3 mg/l

■ 3 - 10 mg/l

● 10 - 20 mg/l

☆ More than 20 mg/l

Source: Mission Eau Nitrates

Figure 6.4 Nitrate in surface waters

through gradually in certain seasons. Nitrates may stay in the soil for 20–40 years before reaching the aquifer, but in the Paris basin the shallow aquifers are often polluted and have been abandoned for drinking water purposes, in favour of deeper ones.

Alluvial aquifers are variable; if they are located in areas with very intensive agricultural holdings they may be vulnerable. This is the case in the valleys of the Garonne and the Rhone, where crops are frequently irrigated. Figure 6.6 shows the results of several research projects on the nitrate content of underground aquifers between 1981 and 1986. The areas most affected are the Paris basin, Bretagne, the Charente and the Garonne valley.

Drinking water

Sixty percent of the drinking water supply in France comes from underground aquifers, while the other 40% is drawn from surface water. In some areas with

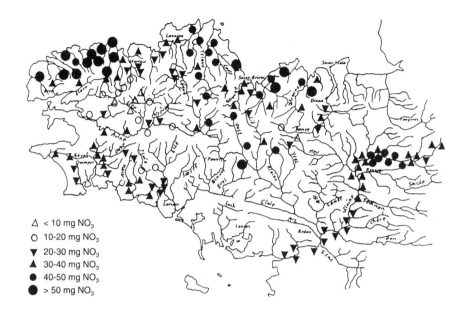

Figure 6.5 Nitrate concentrations in rivers in Brittany, in mg NO₃/l

high levels of nitrate pollution from agriculture the shallower aquifers have to be replaced by deeper ones, because nitrate levels are too high. The drinking water supply is thus kept below the EU safety standard of 50 mg/l in very polluted areas, but this process cannot go on forever. In some areas, occasional high nitrate levels occur because of unusual weather conditions; in these cases, emergency measures have to be taken. The worst affected areas (Fig. 6.7) are:

- Bretagne, because of intensive livestock rearing and the presence of shallow aquifers;
- The Paris basin, because of intensive crop production;
- The south-west (Mid-Pyreneés), with the development of irrigated farmlands, which also give rise to excessive water consumption;
- Also problematic are Nord-Pas-De-Calais, Pays de Loire, Poitou-Charentes and Champagne-Ardenne.

Lakes and marine waters

No specific data are available on the water quality in lakes. In marine water, eutrophication problems exist in the estuaries of the Somme, the Seine and the coastal waters of Bretagne.

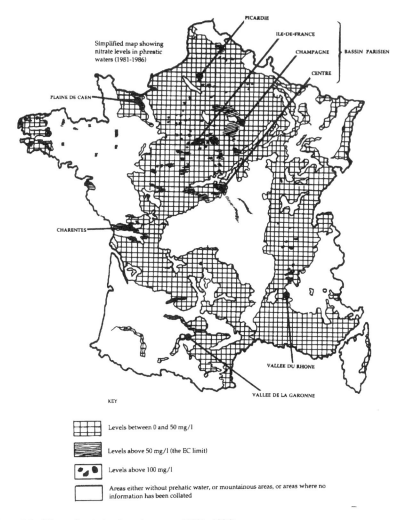

Figure 6.6 Nitrate levels in phreatic waters (1981–1986)

NEED AND PURPOSE OF EXISTING POLICY AND REGULATION (INTERNATIONAL AND NATIONAL)

International policy

Until the early 1980s, little attention was paid to nutrient pollution of ground- and surface water. The INCS conferences and the Paris Conventions, where France was one of the participants, have created more public and political awareness of the problems. France has accepted most of the INSC measures and PARCOM recommendations (see Table 3.4).

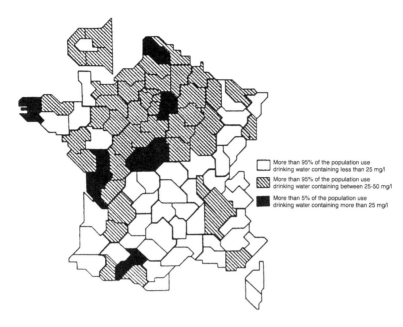

Figure 6.7   Average nitrate content in drinking water (1981)

The status of the EU Nitrate Directive is that legislation for most of the Directive passed in August 1993 and notified to the Commission. A second decree is in preparation. The designation of vulnerable zones is almost complete and 90% of all vulnerable zones have been notified; all envisaged zones together will be about 50% of the agricultural area, or 1/3 of the national territory. The Code of Good Agricultural Practice has been notified.

National policy

The publication of the Henin report in 1980, about the relationship between agricultural activity and water quality, focused public attention on nutrient pollution and started the 'agriculture/environment' debate in France. This report was the result of a working party comprising government departments (Agriculture and Environment), water authorities, agricultural organisations, agronomic research institutes (INRA) and representatives of the fertilizer and pesticide industries.

This working party was set up in 1979 and was asked to gather information on ground- and surface water pollution by agricultural activity, to estimate its extent and to identify means by which the risk of water pollution could be diminished. The report they produced was significant, because it acknowledged for the first time the responsibility of farming for water pollution, which was based on a consensus of the people involved, many of them farmers, about the extent of this responsibility. It also expressed an intention to ameliorate these problems.

Out of this initiative came a recommendation to set up a permanent administrative body to implement the recommended measures and to coordinate the campaign against nitrate pollution: the Steering Committee for the Reduction of Nitrate Pollution in Water or CORPEN. Its executive body is the Mission Eau-Nitrates or MEN. MEN consists of engineers from the Ministry of Agriculture and the Ministry of the Environment, whose job it is to supervise the implementation of measures aimed at combating pollution; they also play an important role in collating and circulating information.

In general, the role of CORPEN and MEN is to promote research studies aimed at perfecting techniques of nitrate pollution control. Efforts have been directed towards improving farming practice and at distinguishing between practices liable to cause nitrate pollution and those that do not affect the water supply. MEN and CORPEN operate on a 'consensus strategy', based on the principle that it is necessary to reconcile high performance in agriculture with water quality, without making any fundamental changes in its development model. This model is still based on growth of productivity and/or intensification of production. Technical adjustments should be perfected and tested and farmers and their advisers could be encouraged to adopt them. Action should be achieved through incentives, rather than through constraining methods, even though these may need to be used later.

POLICY FORMULATION AND LEGAL INCORPORATION

Act on classified installations, 1976

The Act of 1976 on classified installations for environmental protection and its implementation orders supersedes the 1917 Act on hazardous establishments, which did not cover rearing activities. All activities falling under this act have to be classified. For livestock rearing installations, depending on their size: for smaller operations, there is a 'reporting regime', for larger ones a 'permission regime' (Table 6.5). The permission regimes are more demanding, since detailed information has to be supplied on the source, nature and magnitude of disamenities liable to result from the installation. Disamenities include noise, use and discharge of water, the protection of underground water and waste disposal. Requirements for

Table 6.5 Legal regime for classified livestock rearing installations according to species and size

	Reporting regime	Permission regime
Pigs >30 kg	50–450	>450
Poultry	5 000–20 000	>20 000
Rabbits >30 days	2 000–6 000	>6 000
Veals or fattening bovines	50–200	>200
Milk cows	40–80	>80
Nursing cows	>40	>80

manure storage facilities fall under this act. The Act on Classified Installations is a national act; each 'Departement' has its own Departmental Sanitary Regulations, that supplement the national act on a regional level. In regions with a very intensive livestock sector, the regulations are becoming stricter and more geared specifically towards the environmental consequences of livestock farming.

The Water Act of 1964, 1992

The Water Act of 1964 created a management system for the distribution of water and control of water pollution. Under this Act, the Water Authorities of the six major hydrographic networks determine the water quality targets for watercourses. Quality and flow problems are viewed economically, through a system of charges. Pollution discharges are subject to strict standards: a system of charges on water pollution is established, but are limited to point pollution sources. The Water Authorities collect the payment of charges, which are then redistributed through grants, soft loans and rewards for abating pollution. The Water Act of 1992 is a more comprehensive act, which aims to protect the aquatic ecosystems and wetlands, the quality of surface- and groundwater and a better use of water from an economic point of view. Its objective is a balanced management of all water resources through new watershed management schemes.

Implementation of the EU Nitrate Directive

The Decree of 27 August 1993 concerning the protection of waters against the pollution by nitrates of agricultural origin defines vulnerable zones, 90% of which have been notified, and the Code of Good Agricultural Practice, which has been notified. The Decree specifies how to define vulnerable areas at a regional level: the zoning is based on the average level of nitrate concentration during the past year. Waters for which the level of nitrate is over 50 mg/l are automatically considered as polluted waters and consequently they are considered 'vulnerable'. On this basis, about 90% of all vulnerable zones have been notified. Special attention is paid to intermediate areas, where nitrate concentrations are over 25 mg/l, where time series will be set up to discern possible trends. If the trend indicates a nitrate level over 50 mg/l in 2005, the aquifer or river is classified as polluted. In waters where nitrate concentrations reach 40 mg/l and where no time-series exist, pollution is expected to occur in the near future.

Interministerial agreement between the Ministry of the Environment and the Ministry of Agriculture of 11 March 1992

This special agreement was concluded with the objective of introducing a progressive polluter-pays principle into agriculture, as was already the case with other activities. This agreement includes various EU Directives related to water quality and the protection of waters against nitrate pollution from agricultural sources. In this agreement, priority is given to preventive measures, especially advice to farmers, but economic incentives are also included: on the one hand subsidies to encourage farmers to use better methods, reduce run-off and leaching of effluents from manure storage and silage and by the processing of manure and on the other hand levies that will be charged on the basis of a very complex system, that includes number of livestock, condition of buildings, storage facilities, run-off from buildings and manure spreading scheme.

Research and extension programs have been set up by CORPEN and are an integral part of the package. This committee promotes research projects on technical issues and provides advice to farmers; they have also elaborated the codes of good agricultural practice to local conditions.

Decree on the protection of water abstraction points (January 1989)

This decree passed in conformity with Article L20 of the Public Health Act. It stipulates the establishment of three concentric zones for the protection of drinking water abstraction points:

- an immediate perimeter, small in area, where no land use is permitted;
- a close perimeter, average 20 ha, where land use permitted on condition that no pesticides and limited amounts of fertilizer are used; it can be declared out of bounds, on payment of compensation;
- a distant perimeter, average 30 ha, over which constraints are very slight.

Only a small percentage of abstraction points benefit from this protection, because of high costs to rural communities and no compensation is paid to farmers to comply with public health regulations. The exception is the region of Cotes d'Armor, where considerable problems with both eutrophication and nitrate pollution of drinking water have necessitated stringent measures. Here, compliance with regulations concerning drinking water protection zones is being enforced, in combination with advice from two special inspectors and compensations for lost revenue.

Some of the measures that have been taken in Cotes d'Armor on the basis of the Sanitary Regulations (1990) are the following:

- manure storage capacity for 6 months required;
- application of manure prohibited between 15 November and 15 January;
- a fertilizer maximum of 200 kg N/ha for arable land and 350 kg N/ha for grassland, including both manure and mineral fertilizer.

REGULATORY CONTROLS

Even though there has been a legal basis to levy charges on non-point sources since 1975, no levies have been charged so far. In 1982, the water authorities proposed a system of charges per pig according to the number of places available and slurry spreading quality. However, this system never became operable

because agricultural organizations opposed it. In June 1991, the French government decided to introduce a levy on nitrogen, in order to reduce nitrate pollution of ground- and surface water. This levy became part of the interministerial
agreement of March 1992.

Water effluent charges

The water charge system in France has existed from the late 1960s with regard to
point sources. It is now proposed to extend them to livestock farming and cover four
substances: suspended solids, oxidizable matter, reduced nitrogen and phosphates.
In the future, other substances may also be taken into account, e.g. heavy metals and
some organic toxins. Calculating the levy takes three different steps:

1. For each pollutant and each category of livestock (dairy cow, pigs, poultry
   etc.) emission in physical units are calculated. For these calculations, technical
   coefficients, based on industry averages, are used to transform the number of
   animals into quantities of polluting substances. The monetary coefficients are
   used to obtain a gross charge per farm. These monetary coefficients are
   pollutant-specific and do not vary across industries.
2. Farms are classified according to a number of criteria, including manure storage
   facilities capacity, siting of buildings, run-off from buildings, manure spreading
   scheme and livestock density. For each farm an estimate is made of the extent
   to which polluting emissions are reduced and prevented. This estimate is then
   converted into monetary terms, to calculate a premium for the farm.
3. The difference between the gross charge and the premium is obtained: the net
   charge. The charge system and especially the technical coefficients result from
   negotiations between parties and therefore bear the characteristics of a
   compromise. These charges do not include suspended solids or phosphates,
   because a 100% abatement is assumed for these substances. Farmers will have
   to pay a net charge if it is below Ff 6458 (in 1996).

The initial plan was to introduce the charge system gradually, with the following
timetable:

- for rearing installations classified before 1 January 1991, charges will be paid
  from 1 January 1993;
- for rearing installations classified during 1992, charges will be paid from 1
  January 1994;
- for the whole of animal husbandry and crops, the charges will be paid from 1
  January 1995;

For each category subsidies will be given during the year before the levies are due
to help farmers to meet the standards required.

French farmers have protested against the proposed calendar and have also
opposed the principle of the tax itself. Farmers' organizations argue that agriculture
is a special case because of the special circumstances of the agricultural economy.
Pig producers argue that the levies on intensive animal rearing, as expressed in

the environmental classes (1–9) are unfair because mineral fertilizers are not taxed. Also, pork prices have been under pressure because of a large increase in European pig production, with little prospect of prices increasing in the near future, so pig producers' financial position leaves no room for extra levies.

All these factors have convinced the government to delay the introduction of levies. According to the new timetable, the biggest farms will enter the levy system in 1995. Implementing the whole scheme will take 5 years. Moreover, to reduce the effect of the charge on farm income, there is a period of transition during which farmers will only pay a percentage of the amount. Eligible farmers in 1995 will only pay 40% of the total levy, in 2000 they will pay 76%. This process is achieved in 2003, when farmers will pay the full levy.

MONITORING AND CONTROL, RECORD KEEPING AND VERIFICATION

Presently, the only binding legislation is the Act on Classified Installations, which demands that manure storage facilities, housing for livestock and siting thereof, meet certain requirements. However, these requirements are checked only when facilities are first constructed or expanded and little control is exercised thereafter. In Cotes d'Armor, where regulations are much stricter than in the rest of the country, there are also few compliance checks: usually there are about 100 compliance checks a year, versus 800 a year for permit applications. In order to obtain all the necessary information to calculate the proposed levies on effluents, an environmental audit is required. This is a compulsory assessment of the farm and its environmental impact. The system is called DEXEL and is an environmental audit of the livestock farm. This statement consists of two parts:

1. A detailed description of polluting emission flows with emphasis on building siting, storage facilities for manure, run-off, leaching and infiltration.
2. An assessment of agricultural practices whose main outcome is a nitrogen balance both on plot level and on farm level. This focuses on sources of N surplus.

This statement is used to start a negotiation between each eligible farmer and the basin authority, with the objective to achieve individual agreements which include prescriptions that deal with improvement of farm buildings and the promotion of better farming practices. Investments, which are agreed upon, will be subsidized by state, regional and departmental authorities (30%) and the basin agency (also 30%). There are financial and technical commitments around the classification of the farm, to aim at improvements that will lead to a better environmental classification, which in turn will lead to a reduction in the levy amount. Thus before contracting with the agency, farmers have to trade-off between extra costs of compliance through extra investments and the amount of the levy.

It must be noted that the proposed system is very time consuming and thus very expensive. Individual environmental audits and entering into individual negotiations with all eligible farmers will be very costly on the public budget.

Transaction and administration costs will represent a significant part of the total expenditures. The requirement for individual environmental audits is similar to the Danish measure, which required an individual assessment for storage capacity for manure: it soon became clear that it was impossible to enforce this measure, so responsibility for it was transferred to the agricultural organizations.

In France, the emphasis on changing farming practices has always been very much on research, advice and voluntary cooperation from farmers, not on legally binding, enforceable regulations. As far as financial incentives are concerned, the emphasis has been on giving subsidies rather than imposing levies, though this is likely to change in the future. The present balance between farmers and public interest benefits the farmers, but this could change in response to political and environmental pressures around nitrate content in drinking water and eutrophication of surface water.

FINANCIAL MEASURES AND INCENTIVES

Subsidies are available for the following initiatives:

- Investments in manure storage facilities and improved silage facilities to prevent leaching;
- Use of farming methods that are more compatible with the environment;
- Investments in processing of manure;
- In theory, farmers in drinking water protection areas are entitled to compensation for lower yields, as a result of diminished fertilizer use, but in fact this only happens in the Cote d'Armor area.

Levies are not presently being used, but they are proposed for the near future (see above).

EFFECT OF MEASURES

Few data are available on the effect of nutrient-reduction measures, and none are recent. However, PARCOM has published comparisons of applications of manure and the nitrogen and phosphorus surplus between 1985 and 1990 and also on use of mineral fertilizers between 1985 and 1992. These data show that application rates of manure did not change in that period (see Table 3.13) and neither did the nitrogen and phosphorus surplus (see Table 3.15). The application of mineral fertilizer changed between 1985 and 1992 but in the wrong direction: total-N application increased 13.5%, and total-P increased by 19% (see Table 3.14).

Data on the inputs of nutrients from agriculture and expected results of the measures taken or planned (see Table 3.16) show that France gives provisional estimates of a 6% reduction of P between 1985 and 1992 and 17% between 1985 and 1995. For N the estimates are 3% reduction between 1985 and 1992 and a 10% reduction between 1985 and 1995. These figures refer to total amounts for the whole country and not amounts applied per hectare, as the other data do, which makes them difficult to compare. There does seem to be a certain discrepancy between the data given, however.

114

The theoretical effects of measures A1–A5 of the Code of Good Agricultural Practice could have on the quality of surface waters in France shows that in Bretagne, Normandie and the Loire region, these measures could have a moderate effect of around 40 mg/l on the average, while for the rest of the country the effect would be less (around 20 mg/l; Figure 6.8). The theoretical effect of measure A6 on surface water and deep and shallow groundwater is shown in Figures 6.9–6.11. The effects of equilibrium fertilization will be considerable in Bretagne for all water categories, up to 100 mg/l, in adjacent areas it will be slight to moderate (de Cooman *et al.*, 1995).

ROLE OF DIFFERENT ACTORS

The environmental organizations in France do not play a very active role in the debate. Setting up CORPEN in 1984, when the debate just started, has been a preventive move on the part of the French government: it demonstrated a willingness to take the problems seriously and to involve farmers' organizations in acknowledging their responsibility for nutrient pollution of ground- and surface

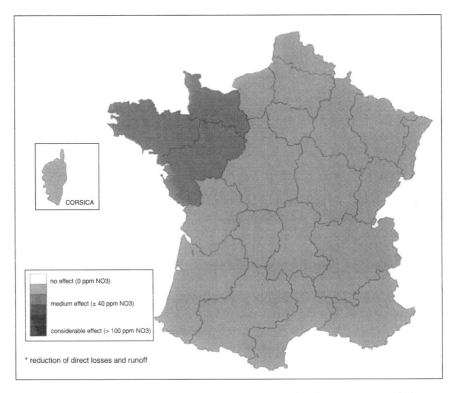

Figure 6.8 Theoretical effect of measures A1–A5 on the quality of surface water. Source: de Cooman *et al.*, 1995

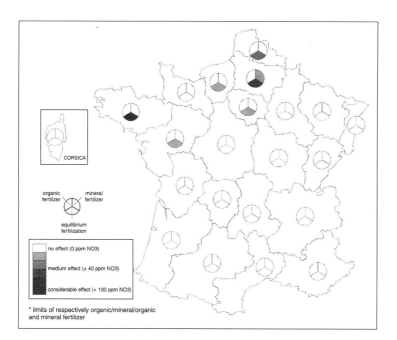

Figure 6.9   Theoretical effect of measure A6 on the quality of surface water. Source: de Cooman *et al.*, 1995

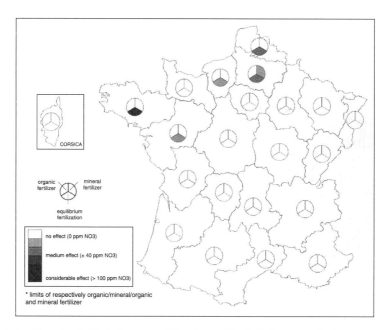

Figure 6.10   Theoretical effect of measure A6 on the quality of shallow groundwater

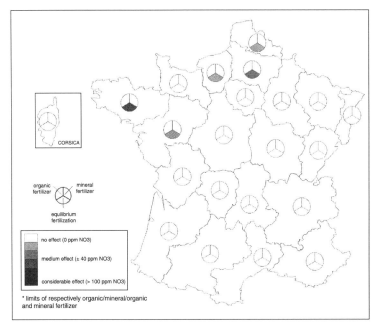

Figure 6.11 Theoretical effect of measure 16 on the quality of deep groundwater. Source: de Cooman *et al.*, 1995

water. On the other hand, doing research and giving advice and information to farmers will probably prove to be insufficient for really solving the problem.

FUTURE DEVELOPMENTS

The future developments have already been touched on above, where a number of compulsory measures and levies, that will come into effect in the next few years, have been discussed.

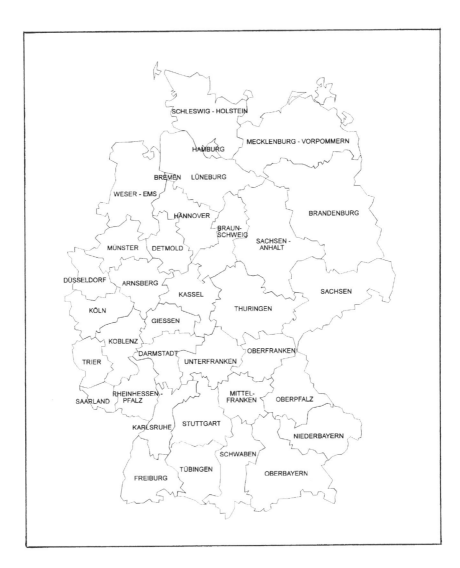

GERMANY

CHAPTER 7: GERMANY

COUNTRY CONDITION

The reunification in 1990 made Germany one of the larger countries in the European Union, with a total area of around 34 million ha, and brought about profound economic and political changes. In the eastern part in particular, conditions have been changing so fast that few statistics at a European level are available. Most of the available data still apply only to the western part of the country.

Germany is a federal republic, consisting of 13 'Länder' (states), that are subdivided in 'Kreise' (counties) and three city-states (Berlin, Bremen and Hamburg). The federal republic defines the general legislative framework, which then has to be implemented in legislation by each of the Länder. This federal structure complicates and slows down the legislative process and laws and regulations will differ among Länder.

The western region (the old Länder)

The agricultural area in the old Länder is about 11.8 million ha, which is about 48% of the total area (23.2 million ha). The agricultural sector consists of around 630 000 holdings and employs 1 million people. The total value of agricultural production amounts to about 29 million ECU. Farms are mostly family enterprises and are an average of 18.7 ha. Smaller farms have decreased in number between 1970 and 1990. In 1987, more than 50% of the agricultural holdings occupied over 30 ha. Of the total agricultural area, 30% is permanent grassland, while 62% consists of arable land. Most of the grassland is situated in the north-western part of the country and in the mountainous area in the middle. In the areas with loess soils mostly cereals and industrial crops are grown. The area devoted to cereals is between 50 and 70% of the arable land (de Cooman *et al.*, 1995).

The old Länder had a livestock population of about 13.2 million LU in 1990, which is 17% of the European livestock population, and comprised 75% cattle, 19% pigs, 3% poultry and 3% other (Table 7.1). Between 1983 and 1990, the livestock population decreased by 3%, mostly cattle, the number of pigs and poultry is fairly stable.

Livestock density is over 2.0 LU/ha in only two counties, Weser-Ems (2.6) and Münster (2.9), while Detmold (1.9) and Düsseldorf (1.8) come close. These counties are part of the Länder Niedersachsen and Nordrhein-Westfalen. Most cattle farms are located in the northwestern part of the country (Lüneburg, Weser-Ems, Münster) and the southwest, while most of the pig farms are located in the northwestern part: Düsseldorf, Arnsberg, Detmold, Hannover and Weser-Ems (de Cooman *et al.*, 1995).

Table 7.1 Structure of livestock population

NUTS II		% of livestock population				
	Total 1000 LU	Cattle (1)	Pigs (2)	Sheep (3)	Horses (4)	Poultry (5)
Rheinhessen-Pfalz	128	52.0	24.7	3.7	3.9	15.7
Braunschweig	236	49.6	42.1	2.2	2.9	3.2
Hamburg	11	70.0	10.9	3.6	14.5	1.8
Karlsruhe	162	60.9	26.5	2.7	4.6	5.4
Unterfranken	290	54.3	39.5	2.2	1.4	2.6
Saarland	64	74.2	13.5	3.3	4.2	4.9
Koblenz	231	64.9	25.2	2.6	2.5	4.9
Freiburg	318	73.2	19.7	1.9	2.3	2.9
Mecklenburg-Vorpommern	1392	55.0	37.9	1.4	0.8	4.9
Darmstadt	218	60.8	25.1	2.4	4.7	7.1
Koln	292	75.4	14.2	2.2	3.9	4.3
Sachsen-Anhalt	1273	47.9	41.5	2.9	0.9	6.8
Brandenburg	1417	51.9	39.3	1.6	0.8	6.4
Oberfranken	337	74.3	22.1	0.8	1.1	1.6
Giessen	212	65.1	27.1	2.7	2.3	2.7
Thueringen	978	54.2	35.6	3.9	0.8	5.5
Trier	211	80.2	15.7	1.8	0.9	1.4
Kassel	384	57.4	36.3	1.6	1.6	3.1
Sachsen	1279	60.6	31.0	2.1	0.7	5.6
Bremen	12	86.1	5.7	0.8	5.7	1.6
Hannover	651	39.3	49.9	0.9	1.4	8.4
Stuttgart	622	56.2	37.0	1.6	1.8	3.5
Oberpfalz	538	77.1	13.7	0.6	0.7	7.9
Luneburg	1070	63.5	30.6	0.7	1.6	3.6
Oberbayern	1133	85.6	10.3	0.8	1.4	1.9
Schleswig-Holstein	1467	70.1	23.9	1.8	1.5	2.7
Tubingen	627	68.9	25.3	1.3	1.5	3.1
Arnsberg	374	52.3	40.0	1.7	2.5	3.5
Mittelfranken	505	68.5	26.9	1.3	0.8	2.4
Neiderbayern	822	61.5	31.5	0.6	0.7	5.7
Schwaben	899	82.7	14.9	0.7	0.8	0.9
Dusseldorf	430	51.8	41.0	1.0	2.8	3.4
Detmold	676	34.3	56.9	0.8	1.1	6.9
Weser-Ems	2604	42.8	42.6	0.3	0.6	13.8
Münster	1212	35.5	60.0	0.3	0.9	3.2

Source: Calculations by de Cooman *et al.*, based on REGlO-Eurostat

The eastern region (the new Länder).

The eastern region consists of 10.8 million ha, 57% of which (6.2 million ha) is agricultural land. About 76% of that area is arable land, only 20% is grassland. Cereals are the most important crop and cover about 50% of the area.

In 1990, the new Länder had a livestock population of about 6.3 million LU, consisting of 54% cattle, 37% pigs, 6% poultry and 3% miscellaneous. Before unification, livestock farming used to be very intensive, with some farms having more than 80 000 pigs each. After unification, a considerable percentage of cattle, pigs and poultry were slaughtered for consumption, which has reduced the density of livestock farming in this part of the country. Presently, the livestock density here is between 0.9 and 1.2 LU/ha.

ENVIRONMENTAL PRESSURES AND SOURCES

Use of mineral fertilizer

In Germany, the average use of fertilizer is 128 kg N/ha (see Table 1.1), varying from a low of 85 kg N/ha for Stuttgart to 175 kg N/ha for Braunschweig (Schleef and Kleinhanss, 1996). The changes over time in the utilization of nitrogen mineral fertilizer are shown in Table 7.2. The use of mineral fertilizer increased steadily between 1970 and 1980, stabilized between 1980 and 1989, and showed a distinct drop in 1990, for the country as a whole and both east and west individually. Table 7.3 shows that the use of mineral fertilizer in Germany exceeds the official advisory norms in about one-third of the regions and exceeds 20 kg N/ha only in three of them (de Cooman *et al.*, 1995).

Use of animal manure

The average production of livestock manure for the whole country is 82 kg N/ha (see Table 1.1). The lowest production level is in Braunschweig (37 kg N/ha), while the highest levels are found in Münster (151 kg N/ha) and Weser-Ems (146 kg N/ha). These two areas, which have a much higher average production than other regions, have the highest livestock densities and highest concentrations of pig farms. Table 7.4 shows that in none of the regions does manure production exceed the limit of 170 kg/ha (de Cooman *et al.*, 1995).

Research at farm level shows that in Germany 12% of the holdings have supply levels of animal manure exceeding 170 kg N/ha: the average manure supply at

Table 7.2 Apparent utilization of mineral fertilizer 1970–1990 (1000 ton N)

	1970	*1975*	*1980*	*1985*	*1986*	*1987*	*1988*	*1989*	*1990*
Germany	1642	1906	2303	2286	2287	2374	2413	2254	1788
West-Germany	1131	1228	1551	1516	1578	1601	1540	1487	1180
East-Germany	511	678	752	770	709	774	873	766	608

Source: OECD

Table 7.3 Mineral fertilizer (kg N/ha per year)

NUTS II	Limit	Utilization	Difference
Bremen	134.9	96.9	−38.0
Schwaben	173.5	146.5	−27.0
Münster	152.8	127.9	−24.9
Weser-Ems	148.2	123.5	−24.7
Düsseldorf	171.8	150.9	−20.9
Trier	152.4	132.5	−19.9
Oberbayern	153.2	134.4	−18.8
Arnsberg	146.5	127.8	−18.7
Lüneburg	139.9	122.2	−17.7
Freiburg	117.8	101.0	−16.8
Köln	175.9	159.3	−16.6
Tübingen	147.5	131.7	−15.8
Detmold	148.3	134.3	−14.0
Saarland	117.7	106.9	−10.8
Oberpfalz	147.6	137.9	−9.7
Mittelfranken	149.3	140.4	−8.9
Schleswig-Holstein	166.9	158.2	−8.7
Brandenburg	92.5	85.6	−6.9
Stuttgart	143.3	137.4	−5.9
Niederbayern	142.7	137.4	−5.3
Koblenz	140.0	136.3	−3.7
Giessen	134.9	132.6	−2.3
Mecklenburg-Vorpommern	111.3	111.5	0.2
Thueringen	118.1	119.3	1.2
Hannover	150.3	152.6	2.3
Oberfranken	130.7	133.4	2.7
Kassel	139.1	142.6	3.5
Sachsen-Anhalt	103.5	107.6	4.1
Karlsruhe	116.9	121.8	4.9
Sachsen	137.1	142.6	5.5
Darmstadt	124.7	131.3	6.6
Hamburg	105.5	119.6	14.1
Unterfranken	126.7	150.2	23.5
Braunschweig	151.1	174.6	23.5
Rheinhessen-Pfalz	101.6	127.3	25.7

Source: de Cooman *et al.*, 1995

these farms is 207 kg N/ha, which, compared to other European countries, is fairly low (see Table 3.3; Brouwer *et al.*, 1995).

Table 7.5 shows manure production and mineral fertilizer use and utilization, compared with the limits defined by the Code, with a utilization coëfficient of 70% for manure. It shows fairly small surpluses in most regions, the largest being in Münster and Weser-Ems, where manure production is high.

Table 7.4 Organic fertilizer (kg N/ha per year)

NUTS II	Limit	Production	Difference
Rheinhessen-Pfalz	166.3	30.0	−136.3
Braunschweig	169.4	37.4	−131.9
Unterfranken	168.5	51.6	−117.0
Karlsruhe	167.8	54.8	−113.0
Mecklenburg-Vorpommern	166.0	57.6	−108.4
Koblenz	168.7	61.4	−107.2
Darmstadt	168.8	62.4	−106.5
Sachsen-Anhalt	163.5	59.3	−104.3
Saarland	168.6	65.1	−103.5
Brandenburg	164.3	61.2	−103.1
Freiburg	166.3	64.4	−101.8
Hannover	169.4	71.9	−97.5
Köln	167.8	70.6	−97.2
Giessen	169.7	72.7	−97.0
Kassel	169.5	72.8	−96.7
Oberfranken	168.6	73.1	−95.5
Thueringen	161.6	69.5	−92.1
Hamburg	150.6	62.0	−88.6
Trier	169.5	83.8	−85.8
Stuttgart	167.6	83.6	−84.0
Lüneburg	167.4	84.5	−82.9
Sachsen	160.5	78.6	−81.9
Oberpfalz	169.2	89.8	−79.5
Arnsberg	167.7	89.4	−78.2
Niederbayern	169.2	92.6	−76.6
Schleswig-Holstein	168.4	92.0	−76.4
Bremen	170.0	96.0	−74.0
Mittelfranken	169.0	95.6	−73.4
Tübingen	166.7	96.5	−70.2
Oberbayern	169.1	100.9	−68.2
Detmold	169.2	104.5	−64.7
Düsseldorf	168.5	113.6	−54.9
Schwaben	169.0	119.4	−49.6
Weser-Ems	169.2	148.1	−21.1
Müster	169.2	154.5	−14.7

Source: de Cooman *et al.*, 1995

Green cover in winter

Climatological and soil conditions vary so much across the country that no federal requirements exist. In some regions such as Baden-Württemberg, there are areas for which a green winter crop is required. In other areas, winter crops are used not only as green fertilizer, but also to prevent erosion on sloping soils.

Table 7.5 Organic and mineral fertilizer (kg N/ha per year)

NUTS II	Limit	Utilization	Difference
Freuburg	117.8	146.1	28.3
Bremen	134.9	164.1	29.2
Köln	175.9	208.7	32.8
Saarland	117.7	152.5	34.8
Brandenburg	92.5	128.5	36.0
Trier	152.4	191.1	38.7
Koblenz	140.0	179.3	39.3
Mecklenburg-Vorpommern	11.3	151.8	40.5
Lüneburg	139.9	181.4	41.4
Karlsruhe	116.9	160.2	43.2
Arnsberg	146.5	190.4	43.9
Sachsen-Anhalt	103.5	149.1	45.6
Rheinhessen-Pfalz	101.6	148.3	46.7
Giessen	134.9	183.5	48.6
Braunschweig	151.1	200.8	49.7
Thueringen	118.1	168.0	49.9
Darmstadt	124.7	175.0	50.2
Tübingen	147.5	199.2	51.8
Oberbayern	153.2	205.0	51.8
Hannover	150.3	202.9	52.6
Stuttgart	143.3	195.9	52.7
Oberpfalz	147.6	200.7	53.2
Oberfranken	130.7	184.6	53.9
Kassel	139.1	193.6	54.5
Schleswig-Holstein	166.9	22.6	55.7
Schwaben	173.5	230.1	56.6
Hamburg	105.5	163.1	57.5
Mittelfranken	149.3	207.3	58.0
Düsseldorf	171.8	230.4	58.6
Detmold	148.3	207.4	59.1
Niederbayern	142.7	202.2	59.5
Unterfranken	126.7	186.3	59.6
Sachsen	137.1	197.6	60.5
Weser-Ems	148.2	227.1	78.9
Münster	152.8	236.1	83.2

Source: de Cooman *et al.*, 1995

Nitrogen balance

The average nitrogen balance for Germany is 65 kg N/ha (see Table 1.3). The averages for regions with surpluses over 100 kg N/ha do not show very large differences, with exceptions for Münster (167 kg N/ha), Weser-Ems (155 kg N/ha), Düsseldorf (146 kg N/ha) and Detmold (138 kg N/ha). The other regions are fairly homogeneous and vary between 102 and 125 kg N/ha. No regions have nitrogen balances exceeding 170 kg N/ha (Schleef and Kleinhans, 1996).

AMBIENT ENVIRONMENTAL CONDITIONS OF SOILS, SURFACE-, GROUND- AND DRINKING WATER

Surface water quality

The most important river basins in Germany are the Rhine and the Elbe. In the Rhine basin in Germany live about 32 million people and it comprises some of the most industrialised areas in the world. Tables 7.6 and 7.7 show nitrate and ammonium concentrations in several German rivers over time. Nitrate and N concentrations increased until 1985/1988, then decreased for 1989/1991. Ammonium concentrations show a steady decrease over the whole period. It is estimated that in Germany 26% of the P input and 45% of the N input into surface and groundwater comes from agricultural sources (BMU, 1993). Due to improved wastewater treatment facilities and the growing use of phosphate-free detergents, it was possible to reduce total surface water pollution by 30% between 1975 and 1987. Pollution originating from agriculture, however, has risen by more than 10%.

Groundwater quality

In Germany, public concern about nitrates has been especially focused on nitrates in groundwater, and the contributing role of agriculture has been acknowledged. A modeling study by several German scientific institutes has estimated nitrate concentrations in ground water under different conditions. Figure 7.1 shows nitrate concentrations of groundwater given maximum detention times. It shows that groundwater in the areas of the Niederrhein, Münsterland, Fränkische Alp, Schwabische Alp, Magdeburg, Thüringen and Maifranken may be threatened by nitrate pollution.

Table 7.6   Quality of river water

River	Nitrate concentration: mg N/l*				
	1970	1975	1980	1985	1991
Rhein/Kleve-Bimmen	1.82	3.02	3.59	4.20	3.60
River	Ammonium concentration: mg N/l*				
	1970	1975	1980	1985	1991
Rhein/Kleve-Bimmen	1.43	1.22	0.59	0.52	0.30
Weser/Intschede	0.60	0.58	0.17	0.09	0.28
Donau/Jochenstein	0.16	0.27	0.16	0.22	0.21

Source: OECD

*Data represent the concentrations on dissolved matter

Table 7.7 Quality of river Data from Eurostat

River	Nitrate concentration: mg NO_3/l*						
	1970	*1975*	*1980*	*1985*	*1987*	*1988*	*1989*
Rhein/Kleve-Brimman	8.06	13.37	15.90	18.60	16.39	16.39	4.10
Weset/Intschede	–	19.04	24.00	22.50	22.23	24.00	5.51
Elbe**	–	–	17.27	13.29	20.81	22.76	–
Donau/Jochenstein	0.89	1.33	2.21	2.66	2.21	2.66	0.50

	Ammonium: mg NH_4/l*						
	1970	*1975*	*1980*	*1985*	*1987*	*1988*	*1989*
Rhein/Kleve-Bimman	1.84	1.57	0.76	0.67	0.45	0.31	0.33
Weser/Inschede	0.77	0.75	0.22	0.14	0.30	0.14	0.14
Elbe**	–	–	1.93	3.86	1.67	0.51	–
Donau/Jochenstein	0.15	0.27	0.15	0.22	0.18	0.14	0.15

* Data represent concentrations in dissolved matter
** 1980, 1985 Geesthacht; 1989 Brunsbüttel

Drinking water quality.

As drinking water supplies are extracted mostly from groundwater (85%), drinking water quality depends greatly on the quality of groundwater. Long-term studies have shown that nitrate concentrations in groundwater have increased in the last 10 years at a rate of 1–1.5 mg/l a year. Especially problematic is the sudden breakthrough of nitrate, which has already been observed in several water extraction works. As a result of this phenomenon, it is possible for nitrate concentrations to increase 10 mg/l in one year, due to the increasing exhaustion of the denitrification capacity of the substratum. Because of the obligation to meet the EU standard of 50 mg/l nitrate in drinking water, this issue will have consequences for an increasing number of drinking water extraction works, that may have to be closed. The development of nitrate concentrations in drinking water since the beginning of this century, is shown in Fig. 7.2.

The Drinking Water Decree of 1986, which was modeled after the EU Directive 80/778 for the quality assurance of drinking water requires a standard of 50 mg NO_3/l and a goal of 25 mg NO_3/l. The concentration of 25–50 mg NO_3/l is reached in 710 Bayern drinking water points representing 16% of the total withdrawal, while 350 points representing 5.7% of the withdrawal exceeded the 50 mg/l at least once. A national survey showed that 22.4% of the groundwater stations in Niedersachsen exceeded 50 mg NO_3/l followed by 16% in Nordrhein-Westfalen. In the median state 7% of the samples exceeded 50 mg NO_3/l (Fig. 7.3).

Individual wells have shown substantial increases over time. Two of the wells

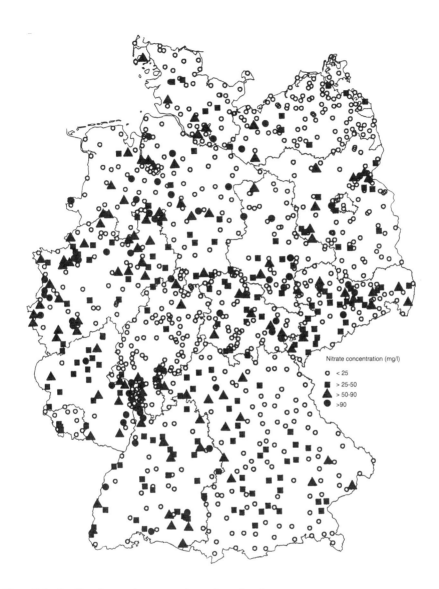

Figure 7.1 Quality of groundwater: maximum detention time

of the City of Bruchsal (2.4 million m³/year for 23 000 inhabitants) in Baden-Wurttemburg experienced increases of 15–30 mg/l/year during a 2-year period, after which the level stabilized. The well with the highest N level withdrew from agricultural land, while the well with the lowest level had only forest in its infiltration area. The well from the adjacent Neuthard-Karlsdof (0.4 million m³/year for

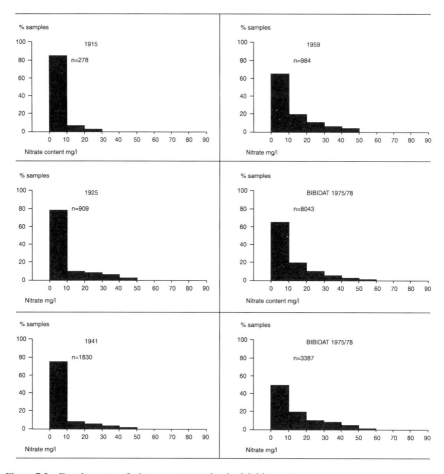

Figure 7.2   Development of nitrate concentration in drinking water

8 000 inhabitants) located 300 meters further, increased at a rate of 10 mg/l/year. This well also withdrew from farmland (Fig. 7.4). The subsequent stabilisation and decrease is likely due to subsequent initiation of denitrification (MUBW, 1989)

Lake and marine water quality
The water quality of the Bodensee showed a steady increase in N-total between 1970 and 1990, from 0.73 to 0.98 mg N/l. The quality of marine water is very much threatened by eutrophication. As Fig. 3.4 shows, the Deutsche Bucht in particular has been classified as vulnerable to nutrient pollution. Here, high concentrations of nitrates can be observed at distances up to 50 or even 100 km

	<1*	>1–10*	>10–25*	>25–50*	>50–90*	>90*
Baden-Württemberg	7,0	23,0	30,0	29,0	9,0	2,0
Bayern	20,0	29,8	22,5	5,7	0,8	
Berlin	71,0	19,2	4,3	3,3	1,1	1,1
Brandenburg	47,1	24,6	18,0	6,4	1,3	
Bremen	51,0	29,0	8,0	3,0	7,0	2,0
Hamburg	61,7	11,2	6,5	13,1	7,5	–
Hessen	32,0	35,0	18,0	7,5	4,0	3,5
Mecklenburg-Vorpommern	68,0	18,0	5,0	2,0	7,0	–
Niedersachsen	51,3	14,3	5,8	6,2	8,9	13,5
Nordrhein-Westfalen	18,0	23,0	21,0	22,0	9,0	7,0
Rheinland-Pfalz	38,1	29,0	11,5	8,9	6,7	6,1
Saarland	10,9	30,0	32,7	23,6	2,7	–
Sachsen	5,8	24,4	23,3	30,2	11,5	4,7
Sachsen-Anhalt	33,3	35,2	1,8	14,8	5,6	9,3
Schleswig-Holstein	45,0	22,0	12,0	10,0	7,0	4,0
Thüringen	15,8	25,1	22,5	21,9	9,7	4,9
Min.	6	11	2	2	1	0
Max.	71	35	33	30	12	14
Median	36	25	12	14	7	3
Average	36	25	14	15	7	4

Source: Landesarbeitsgemeinschaft Wassser, 1995

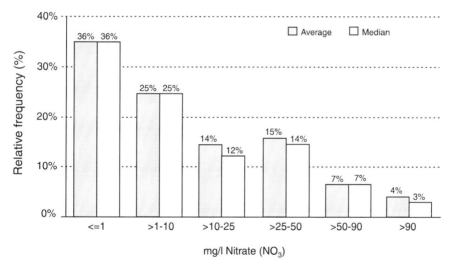

Figure 7.3 Nitrate concentrations in groundwater in Germany. Frequency distribution of nitrate contents (*in mg/l NO_3; data for Bundesländer in %)

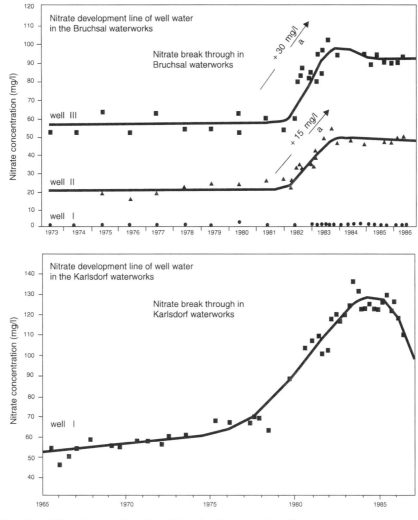

Figure 7.4 Nitrate increases in wells withdrawing from agricultural land

from the coast. Long-term studies show that between 1962 and 1985 the concentrations of nitrogen in coastal waters almost doubled.

NEED AND PURPOSE OF EXISTING POLICY AND REGULATION (INTERNATIONAL AND NATIONAL)

International policy

Germany has been confronted with the problems of nutrient pollution relatively early, compared to other European countries, because the eutrophication of the

North Sea in the Deutsche Bucht started to manifest itself in the late 1970s. These problems became even more prominent and visible between 1983 and 1988. As was mentioned in Chapter 2, a German report on the water quality of the North Sea (Umweltprobleme der Nordsee, published by Der Rat von Sachverständigen für Umweltfragen) was published in 1980 and became the stimulus for the first INSC in Bremen in 1984. Another contributing factor has been the events in Chernobyl in 1986, after which a new Ministry was created for environmental policy, the Bundesministerium für Umwelt, Naturschutz and Reaktorsicherheit, which has strengthened the position of this policy area. Until then the environment was part of the Ministry for the Interior. As initiator of the first INSC conference, Germany has accepted all decisions of the first INSC, and most decisions of the later conferences (see Table 3.4). It has also accepted most Parcom recommendations.

The status of the Nitrate Directive is that the whole area has been designated as vulnerable zone, the Code of Good Agricultural Practice has become part of the new Fertiliser Decree (Düngeverordnung) of January 1996. It is also the intention to present the Fertiliser Decree as the first Action Program to the Commission.

National policy

In Germany, environmental organizations have been influential for a long time. The Green political party has been very successful and the media have paid considerable attention to environmental issues. Initially, most of the concern focused on eutrophication of surface water, notably the North Sea, but this focus has now shifted. Presently, the most urgent water pollution problem caused by agriculture is the (already high and still rising) nitrate pollution of groundwater, which was not as 'visible' in the early 1980s, but has become more prominent as political concern and in the public awareness since then. The eutrophication of surface water by increasing N and P influxes still rates a close second priority.

POLICY FORMULATION AND LEGAL INCORPORATION

Water Management Act (Wasserhaushaltsgesetz – WHG) 1957, amended in 1986 and 1996

Water, like air, is regarded as an indispensible natural resource and as such subject to special legal protection. The WHG applies to surface water, coastal waters and ground water. The conditions of liability are formulated as follows: "Whoever inserts or introduces substances into water or whosoever influences water in such a way that the physical, chemical or biological composition is changed, is under the obligation to compensate any damaging result for another party." Under this law, water protection areas can be established in order to protect the water supply from harmful influences, particularly the protection of groundwater, which is important for drinking water extraction. When the law was amended in 1986, a special protection measure for the surface water was added, because of the increased risk of eutrophication.

Water Protection Areas consist of three different zones, from small areas close to the extraction point, usually owned by the Water Company, to a larger one, within which it should take groundwater 50 days to reach the extraction point, to an even wider one, which should include the whole catchment area. The designation of such areas take a formal procedure, carried out by the district magistrate who is the authority for water at the local level. In 1988, 1.9 million ha (7.6% of the total land area of the Federal Republic) was recognized as such.

When the WHG was amended in 1986 the section that refers to water protection areas was amended to include an entitlement to compensation for 'orderly agriculture' to be made for economic disadvantages caused by these regulations. The problem in this text, however, was that there was no uniform legal definition of 'orderly agriculture'. This problem has been rectified by the Fertilizer Decree of 1996, where 'orderly agriculture' has been defined in the Code of Good Agricultural Practice.

The question of how to finance payments to farmers made to reduce fertilization and to have less intensive production in water-protected areas has had a number of diverse solutions in different Länder: in Baden-Württemberg, the 'water penny' (an extra user charge on the water bill to finance compensations to agriculture) has led to a lot of controversy. The fact that Länder have so much discretion to set their own levels of compensation has led to strong controversies among farmers: Länder use extremely divergent compensation rules, giving rise to questions of social injustice.

Waste Water Charges Act (Abwassergabengesetz) 1986

The Waste Water Charges Act supplements the WHG. Because of the waste water charge, which is calculated according to the actual discharge of pollutants, there is an incentive for improvements in water quality. This only applies to point sources, however. For households, a flat rate is charged. There are some problems in the application of this law to agriculture.

Waste Disposal Act (Abfallbeseitigungsversetz)

Article 15 of this law applies especially to agriculture. It specifies that "wastewater, sewage, faeces or similar substances must comply with regulations in this law" if they are deposited on soils for agriculture, forestry or gardening. Cesspools, liquid manure and livestock manure are exempt, if the "normal proportion of agricultural fertilizer use" is not exceeded.

Fertilizer Decree, 1996

The Fertilizer Decree (Düngeverordnung) of the Federal Republic of Germany was approved by the Bundesrat and proclaimed by the Minister of Agriculture (BMELF) on January 26, 1996. The general regulations start on February 6, 1996 while the special provisions are in force as of July 1, 1996. This decree enlarges the Fertilizer Law (Düngemittelgesetz) of 1977 as amended in 1989 and 1994. It codifies at the federal level the Code of good agricultural practice for fertilizer handling and application and covers part of the EU Nitrate Directive (BMELF, 1996). It is the purpose of the decree to provide a better use of fertilizers and to reduce nutrient losses to ecosystems and waters. It supplants the fertilizer regulation that several states had enacted on the basis of the waste laws. The decree covers agricultural and horticulture practices but not private vegetable gardens, golf links or greenhouses.

REGULATORY CONTROLS

The Fertilizer Decree
The Code of Good Agricultural Practice of the Fertilizer decree requires the utilization of nutrients to the fullest extent and with the lowest possible losses. Nitrogen fertilizers can only be applied during the crop growth season and when the soil can absorb the fertilizer. N addition after the harvest in the fall is generally not allowed, unless the next culture has an actual N need, or for humification of crop residue for which 40 kg NH_3-N/ha or 80 kg total N/ha is allowed. Application of liquid manure is prohibited between 15 November and 15 January. The fertilizer need is a function of the crop requirement, the soil nutrients, the calcium and pH and cultivation practice (tilling, watering etc.). Manure is treated similar to mineral fertilizer applications as far as N, P, K application calculations. On the basis of the EU Nitrate Directive, the manure application is limited to 210 kg N/ha for grassland and 170 kg/ha for arable land on the basis of total land size. The application equipment must be state of the art and be able to apply and spread the correct amount with minimal ammonia losses, not to exceed 20% for manure applications. Fertilizers cannot enter water bodies or be applied in adjacent areas. Losses of ammonia are minimized by application at lower temperature, under cloudy skies, low wind velocity, dilution with water, nitrification inhibitors and improved application equipment. Inevitable and allowable N losses during storage are 10% for liquid manure and 25% for dry manure.

Enterprises with more than 10 ha of agriculture and more than 1 ha of horticulture have to provide mineral balance accounts and save these for 9 years.

MONITORING AND CONTROL, RECORD KEEPING AND VERIFICATION

Monitoring and control of the Fertilizer decree will become the responsibility of the Länder. Since the Decree is so recent, no legal incorporation by individual Länder has yet taken place. The Fertilizer Decree requires a mineral balance account to be saved for 9 years. Control and enforcement of these will also be the task of the Länder.

FINANCIAL MEASURES AND INCENTIVES

Financial compensation is offered in Water Protection Areas. The amounts offered vary per Länder (Table 7.8). Possible levies on nitrogen surpluses from fertilizers will also depend on legal incorporation by individual Länder.

EFFECT OF MEASURES

PARCOM data show that the average application rates of manure in Germany have changed between 1985 and 1992: application of total N was 73.5 and increased to 86.3 kg N/ha, while total P decreased from 22.3 to 16.1 kg P/ha in that period (see Table 3.13). Mineral fertilizer application rates decreased between 1985 and 1993 by 25% for total-N and 58% for total-P (see Table 3.14). The nitrogen surplus was reduced by 7% and the phosphorus surplus by 67% between 1985 and 1992 (see Table 3.15). The total inputs of nutrients from agriculture and the expected results of the measures by 1995 show an expected reduction for P of 21% in 1995, and of 17% for N (see Table 3.16).

A modeling study by Schleef (1996) compared the possible effects of the Nitrate Directive and the Fertilizer Decree (based on N, or on NPK) with the impact of CAP Reform. It showed that for the 15% of the farms that are affected by the Fertilizer Decree, there will be a substantial reduction of nitrogen surpluses (Table 7.9) under all conditions. Results from the optimization model indicate that after the introduction of these policies, nitrogen surpluses on these farms will still exceed 180 kg/ha. According to the author "even if all kinds of improvements in manure management are taken into account, a nitrogen surplus

Table 7.8 Compensatory payments in water protection areas in the individual länder (situation in May 1989)

Land	Changes in water laws	Payment	Amount
Baden-Württemberg	27.07.87	Land	DM 310/ha (lump sum)
Bavaria	10.12.87	Water supply works. Land in case of new protection areas	Full compensation (precise estimate)
Hamburg	09.10.86	Water supply works	Full compensation (lump sum possible)
Hesse	July 88 (draft)	Water supply works	Differentiated lump sum
Berlin	Nov 88 (draft)	Water supply works	Not yet regulated
North Rhine-Westphalen	14.03.89	Water supply works	Not regulated
Rhineland-Pfalz	pre-draft	Water supply works	Not regulated
Saarland		Water supply works	Not regulated
Schleswig-Holstein	no changes		
Bremen	no changes		

Source: FIP Bulletin 5/89, p, 118

Table 7.9 Impact of the Nitrate Directive and the Fertilizer Decree by regions and farm size: sub-sample

Measure	Unit	Schleswig-Holstein, Lower Saxony, North Rhine-Westphalia				Hesse, Rhineland Palatinate, Bavaria, Baden-Württemberg			
		CAP reform	change compared to CAP reform in %			CAP reform	change compared to CAP reform in %		
		absolute values	EU nitrate directive	fertilizer decree (N)	fertilizer decree (NPK)	absolute values	EU nitrate directive	fertilizer decree (N)	fertilizer decree (NKP)
No. of farms		51				53			
Gross margin	DM	127 169	-5.08	-0.59	-7.27	89.241	-3.33	-0.02	-1.33
Area	ha	31.76	–	–	–	21.70	–	–	–
Livestock composition									
total livestock	LU	72.29	-16.95	-2.17	-15.42	46.67	-10.65	-0.04	-2.76
cattle	LU	49.31	-16.16	-1.24	-12.67	42.36	-11.71	-3.02	–
pig and poultry	LU	22.98	-18.58	-4.22	-21.32	4.31	-0.23	-0.23	-0.23
Mineral balances									
–nitrogen balance									
min. N input	kg/ha	153.22	2.79	-0.17	17.15	184.78	0.04	-0.01	9.05
org. N input	kg/ha	202.10	-18.60	-2.52	-13.56	193.40	-12.46	-0.05	-2.64
total N input	kg/ha	355.32	-9.38	-1.51	-0.32	378.18	-6.35	-0.03	3.08
N uptake	kg/ha	140.02	-0.88	-0.04	-1.87	174.78	-1.10	0.05	0.02
nitrogen surplus	kg/ha	215.30	-14.91	-2.46	0.70	203.40	-10.87	-0.10	5.70
–phosphate surplus	kg/ha	87.50	-9.03	-2.51	1.14	78.80	-4.88	-0.26	6.20
–potassium surplus	kg/ha	99.40	-9.96	-1.91	2.01	81.30	-5.17	0.62	7.13

–,no change Schleef, 1996

136

Table 7.10 Nitrogen balances under the Nitrate Directive and Fertilizer Decree: comparison between optimization model results (opt.) and the technological approach in (t.a) taking into account only technological potential

Measure	EU nitrate directive		Fertilizer decree (only)		Fertilizer decree (NKP)	
	opt.	t.a.	opt.	t.a.	opt.	t.a.
Min. N input (kg/ha)	168.9	117.8	166.2	108.5	188.7	114.2
Org. N input (kg/ha)	166.6	166.6	195.4	195.4	180.4	180.4
Total N input (kg/ha)	335.5	284.4	361.6	303.9	369.9	294.6
N uptake (kg/ha)	133.1	153.1	154.4	154.4	153.0	153.0
Nitrogen balance (kg/ha)	182.4	131.3	207.2	149.5	216.1	141.6

Schleef, 1996

of more than 130 kg/ha remains" (Schleef, 1996, p.15). Table 7.10 shows the results of this optimisation study under different assumptions.

The theoretical impact of measures A1–A5 (run-off measures) of the code on the quality of surface water show moderate effects for some regions in the northwestern part and in the east and south-east (Fig 7.5). The theoretical effects

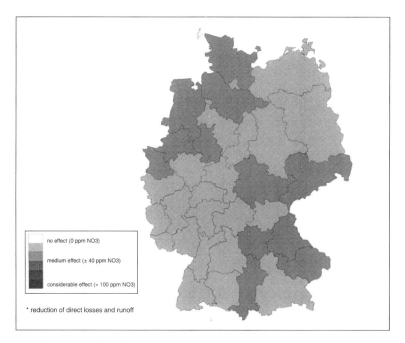

Figure 7.5 Theoretical effect of measures A1–A5 on the quality of surface water. Source: de Cooman *et al.*, 1995

of measure 6 on the quality of surface water and shallow ground water, show considerable improvement due to equilibrium fertilization in those same areas, moderate effects in the rest of the country and moderate effects due to reduced fertilizer use (Figs 7.6, 7.7). The effect on deep ground water is expected to be moderate, except in the northwest and east, where equilibrium fertilization could have an important effect (Fig. 7.8, de Cooman *et al.*, 1995)

ROLE OF DIFFERENT ACTORS

In Germany, the fact that individual Länder have a lot of legislative power makes them important actors. In the past, several Länder have implemented strict legislation to deal with nutrient pollution problems, but there is a certain amount of arbitrariness inherent in the differences between these laws which is hard on farmers: it runs counter to their feelings of social justice. The federal government has been slow in arriving at a new legislative framework, to comply with the requirements of the EU Nitrate Directive. The environmental organizations had a lot of influence in the 1980s in making the public aware of the dangers of nutrient pollution of ground- and surface water and have given the impetus for the international debate that took place.

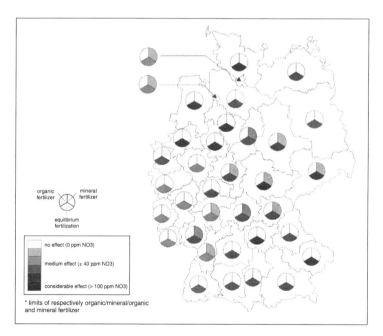

Figure 7.6 Theoretical effect of measure A6 on the quality of surface water. Source: de Cooman *et al.*, 1995

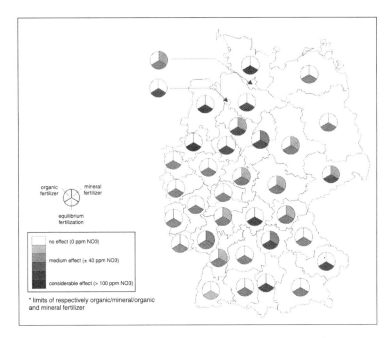

Figure 7.7 Theoretical effect of measure A6 on the quality of shallow groundwater

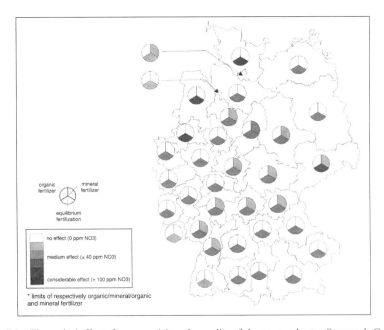

Figure 7.8 Theoretical effect of measure A6 on the quality of deep groundwater. Source: de Cooman *et al.*, 1995

FUTURE DEVELOPMENTS

Since the Fertilizer Decree is so recent, there are no plans yet for further developments. This will also depend on the question whether the Commission will accept the Fertilizer Decree as Action Plan, or make additional requirements.

UNITED KINGDOM

CHAPTER 8: UNITED KINGDOM

COUNTRY CONDITION

The United Kingdom covers about 23.5 million ha, 78% of which is agricultural land (about 1.4 million ha). The agricultural sector consists of about 234 900 holdings and occupies around 533 000 people. The value of its agricultural production is a little over 24 million ECU. Farms are generally large, the average size being 72.4 ha. The number of holdings is fairly stable, but has been reduced by about 2.5% in the period 1990–1995. There is a tendency towards specialization and integration with agro-business firms (de Cooman *et al.*, 1995).

The total agricultural area fell by about 2% per year in the period 1984–1995 and currently consists of 34.7% arable land (including set-aside), the remaining 65.3% being grassland. There are marked differences between the different regions.

- Permanent grassland covers 70% of Scotland, Northern Ireland and the north-west of England. Cereals are the most important crop in this area, except for Northern Ireland, where fodder crops predominate.
- The midlands have mostly mixed farms. Towards the west, there are more fodder crops and grasslands are increasing.
- East Anglia and the south-east have mostly arable farms, growing wheat, barley, sugar beet and potatoes.
- Wales and the south-west have a lot of grassland, but cereals and fodder crops are also important.

The United Kingdom has a livestock population of about 16 million LU, which is 16% of the European livestock population. The 1995 census data for the UK show there were 11.7 million cattle, 42.7 million sheep, 7.5 million pigs and 125.9 million poultry. Compared with 1984, these data show a 9% decrease for cattle, a 19% increase for sheep, a 4% decrease for pigs and a 5% increase for poultry.

Pigs are mostly held in Yorkshire and Humberside, and East Anglia, where they represent 33 and 55% of the livestock population. Scotland, Wales and the south-west all have a livestock population of more than 2 million, of which 60% is cattle. Sheep are also important in Scotland and Wales. Nowhere in the UK does the livestock density exceed 2 LU/ha. The highest densities are found in Northern Ireland and the north-west, but generally the density is fairly homogeneous across the whole country at 1.3–1.8 LU/ha (de Cooman *et al.*, 1995).

ENVIRONMENTAL PRESSURES AND SOURCES

Use of mineral fertilizer
The average use of mineral fertilizer in 1995 in the UK was 130 kg N/ha. There is a difference in the use of straight and compound nitrogen fertilizer: in crop farming, the use of compounds decreased 50%, but utilization of straights increased from 40 kg/ha to 120 kg/ha between 1970 and 1988. Since 1985, the use of mineral fertilizer in crop farming has stabilized. For grasslands, use of both kinds increased for a number of years, but from the early 1990s, utilization has decreased somewhat (Tables 8.1, 8.2).

A comparison between the use of mineral fertilizer with the EU limit in different parts of the UK shows that only in East Anglia exceeds the use of mineral fertilizer the limit, and not very much (Table 8.3; de Cooman *et al.*, 1995).

Use of animal manure
The production of manure in the UK varies per region, the highest levels being found in north-west England (132 kg N/ha) and Northern Ireland (124 kg N/ha). In Scotland it is no more than 44 kg N/ha. Table 8.4 shows that the average level per region nowhere reaches the EU limit of 170 kg N/ha. According to Table 1.3, the average manure production for the whole country is 66 kg N/ha. However, data at farm level show that for the 17% of the holdings with a manure production over 170 kg N/ha, the average manure supply is 258 kg N/ha (see Table 3.3; Brouwer, 1995).

The combined use of mineral fertilizer and animal manure is given in Table 8.5. Again East Anglia exceeds the limit by 55.7 kg/ha, the east midlands does so by 14.1 kg/ha. All other areas remain well below the limit.

Table 8.1   Utilisation of mineral fertilizer (kg N/ha) in straight form

Year	Arable land		Grassland		Total	
	England, Wales	Scotland	England, Wales	Scotland	England, Wales	Scotland
1987	136	82	75	39	105	56
1988	125	74	61	52	94	61
1989	130	74	70	36	100	48
1990	131	82	68	38	100	55
1992	132	74	55	36	94	49
1993	118	80	63	33	89	49
1994	127	72	54	37	90	50
1995	130	90	57	33	92	54

Table 8.2 Utilisation of mineral fertilizer (kg N/ha) in composite form

Year	Arable land		Grassland		Total	
	England, Wales	Scotland	England, Wales	Scotland	England, Wales	Scotland
1987	25	57	58	77	41	69
1988	24	51	55	80	39	68
1989	20	54	57	81	39	70
1990	18	49	64	78	41	67
1991	16	53	64	75	39	67
1992	15	51	49	75	32	67
1993	19	50	49	81	35	70
1994	22	56	63	74	43	68
1995	20	51	63	81	42	70

Table 8.3 Utilization of mineral fertilization (kg N/ha)

NUTS II	Limit	Utilization	Difference
Scotland	329.7	53.7	−276.0
Wales	337.9	83.8	−254.1
Northern Ireland	305.0	88.8	−216.2
North-West	292.2	112.3	−179.9
South-West	262.5	106.8	−155.7
West Midlands	235.8	117.4	−118.4
Yorkshire and Humberside	229.3	127.3	−102.0
South-East	193.2	117.4	−75.8
East Midlands	194.6	151.7	−42.9
East Anglia	148.9	167.8	18.9

Source: de Cooman *et al.* 1995

Green cover in winter

There are no exact data for the use of green manure and cover crops. Winter cereal crops increased from 1970 to 1990 but due to the effects of set-aside they have diminished slightly in the last few years. The area in any single year can be affected by weather conditions.

Nitrogen balance

The nitrogen balance for the country as a whole is 80 kg N/ha. In the middle of the country the nitrogen balances are highest, with Cheshire 117 kg/ha and Humberside 109 kg/ha (see Table 1.3; Schleef and Kleinhanss, 1996). Cheshire is a county with an intensive dairy sector with relatively high stocking rates and fertilizer use. Humberside has an accumulation of pigs in proximity to arable land.

Table 8.4 Utilization of organic fertilizer (kg N/ha)

NUTS II	Limit	Production	Difference
Scotland	167.9	41.8	−126.0
East Anglia	148.7	36.9	−11.8
East Midlands	155.9	57.0	−98.9
South-East	152.7	55.0	−97.7
Yorkshire and Humberside	161.6	81.3	−80.4
North	167.1	108.2	−58.9
West Midlands	158.4	105.5	−52.8
South-West	163.9	11.3	−52.6
Wales	168.5	121.0	−47.5
Northern Ireland	167.1	124.0	−43.1
North-West	163.6	132.9	−30.7

Source: de Cooman *et al.* 1995

Table 8.5 Utilization of organic and mineral fertilizer (kg N/ha)

NUTS II	Limit	Utilization	Difference
Scotland	329.7	95.5	−234.2
Wales	337.9	204.8	−133.1
North	306.0	199.2	−106.8
Northern Ireland	305.0	212.8	−92.2
North-West	292.2	245.2	−47.0
South-West	262.5	218.1	44.3
South-East	193.2	172.4	−20.8
Yorkshire and Humberside	229.3	208.6	−20.7
West Midlands	235.8	22.9	−12.9
East Midlands	194.6	208.7	14.1
East Anglia	148.9	204.7	55.7

Source: de Cooman *et al.* 1995

The positive nitrogen balances are the result of a combination of stock numbers and fertilizer use.

Regional averages can hide large individual differences between farms. As Table 3.3 shows, intensive livestock farms in Britain have high levels of manure production. Many of these are located in the wet, western part of the country which is mainly devoted to grassland and with more than one livestock enterprise. In eastern England there are intensive pig and poultry units where manure loading rates are high.

AMBIENT ENVIRONMENTAL CONDITIONS OF SOILS, SURFACE-, GROUND- AND DRINKING WATER

Surface water

The concentration of nitrates in surface waters such as rivers increased steadily during the period 1970–1990. During the early 1980s the nitrate concentrations increased considerably (Table 8.6). After 1985 nitrate levels seem to stabilize. The Thames and Severn rivers carry a much higher nitrate load than the Clyde and the Mersey; this is not only due to nutrient loss in agriculture, but is also related to domestic waste from sewage treatment plants and natural circumstances. The most important sources of phosphorus are sewage treatment plants and untreated sewage (according to OSPAR data, in 1992 about 17% of domestic wastewater in Britain was being discharged directly into estuaries or coastal waters with no or only primary treatment). There is currently a major investment program in place to improve this situation and meet EU legislation.

Groundwater

In England, about 70% of the drinking water supply is abstracted from surface water. Groundwater sources are to be found mostly in the drier eastern and southern regions of the country. In this region especially, a significant part of the aquifers are vulnerable to nitrate leaching for geological reasons (Figure 8.1) and the low rainfall provides a low dilution factor. This part of the country is also the area that is most favourable for intensive agricultural and horticultural production. Consequently, arable farming predominates over many of the vulnerable aquifers and is mainly responsible for the presence of nitrate in groundwaters, that are utilized for public water supplies.

For groundwater, it is important to make a distinction between aquifers under sandstone and chalk, where nitrate can remain for a long time in the strata above the water table and leach through only gradually (between 20 and 40 years), and aquifers under fractured limestone, where nitrates remain for a shorter period (a number of years). Figure 8.2 shows the fluctuations in the nitrate concentrations in both these types of aquifers.

Table 8.6 Water quality in rivers, nitrate concentrations: mg N/l

River	1975	1980	1985	1990
Thames	6.50	6.90	7.98	8.17
Severn	5.52	5.80	6.33	6.28
Clyde	2.65	1.85	2.15	2.10
Mersey	1.87	2.28	3.12	3.27

Source: OECD

Outcrop with permeable soils (aquifer unconfined)
and directly vulnerable to nitrate leaching

• Boreholes over 50 mg/l NO₃ on one or more occasions in 1983 or 1984

Approximate extent of area in which the nitrate concentration in
unconfined aquifers is likely generally to exceed 100 mg/l in the long term

Source: Adas

Figure 8.1 Outcrops of principal aquifers and the vulnerability of groundwater to nitrate pollution

Drinking water

In 1987, about 10% of the catchment areas for drinking water presented water quality problems. In 25% of these cases, the problems were related to nitrate levels of ground water with a nitrate content exceeding the limit of 50 mg/l. These problems tend to be most severe in the south-eastern half of England. In the East Anglia and Severn Trent region, 15 sources of drinking water were permanently closed. Surface waters often have high nitrate concentrations as well, particularly during the autumn and winter leaching period. The concentrations at abstraction points can often be kept below the 50 mg/l limit because of the natural denitrification that occurs in reservoirs, which are widely used for storage purposes

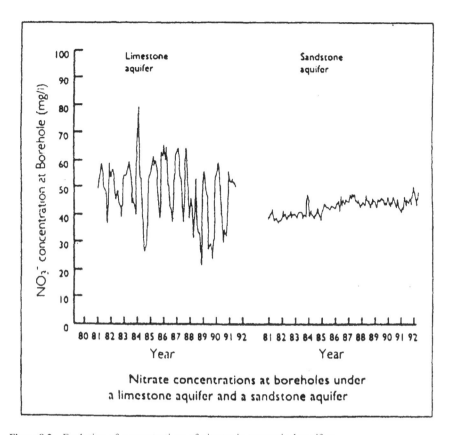

Figure 8.2 Evolution of concentrations of nitrates in two typical aquifers

by water suppliers. Nevertheless, rising nitrate levels in surface water were also a source of concern, both in their own right and because they reduce the availability of river water for blending with contaminated groundwater.

England differs from other European countries, as far as drinking water problems are concerned, in that high nitrate levels are here more related to high mineral fertilizer use, rather than livestock manure, especially in the south-eastern part of the country. Livestock farming is concentrated in the western part, where groundwater is little used for drinking water and dilution is greater so that contamination is less of a concern.

Lakes and marine waters
Lake water quality is variable, with some lakes becoming more polluted with nitrates over time, others less so (Table 8.7). The quality of coastal and marine waters in the UK has not been a problem; as Figure 3.4 shows, eutrophication problems in the North Sea occur along the Danish, Dutch and Belgian coast,

Table 8.7 Lake water quality, N-total, mg N/l

Lake	1980	1985	1990
Neagh	0.48	0.48	0.77
Lomond	0.30	0.29	0.20

Source: OECD

even though the estuaries of the Wash, Ythan and Langstone Harbor may begin to show signs of eutrophication. The prevailing currents in the North Sea favour the UK by diluting potentially polluting substances. However, due to the flow and circulation patterns in the North Sea and the large riverine input from continental rivers, the UK contribution is relatively small.

NEED AND PURPOSE OF EXISTING POLICY AND REGULATION (INTERNATIONAL AND NATIONAL)

In the UK, nitrate is perceived as being mostly a drinking water problem. As indicated above, eutrophication of rivers and coastal waters is not a major problem in Britain, even though increasing nitrate levels in surface water are seen as problems for drinking water supplies.

The UK also uses a different set of standards regarding the control of pollution from point source discharges of non-agricultural nutrients and other substances than the other European countries do. It utilizes environmental quality objectives (EQO): the quantity of a substance that is acceptable in the environment is determined by the 'use function' of that part of the environment. This standard is more advantageous for British industries, since it allows them to discharge large quantities of waste substances in the North sea, where these substances will be diluted and carried away by the prevailing currents, without really exceeding local EQS (environmental quality standard). Other North sea countries use uniform emission standards (UES), which means that each source is not allowed to discharge more than a specified concentration of a certain substance, depending on the best available technology to reduce emissions. For British industry, it would be very costly to switch to UES standards (Vosmer, 1995).

The standard for nitrate in drinking water is laid down in the EC Directive relating to the quality of water intended for human consumption and has been central to the debate in the UK for the last decade. The British government has argued that the EU standard for nitrates in drinking water of 50 mg/l is unnecessarily strict and that the previous British standard of 100 mg/l was acceptable. The government asked for the advice of the Chief Medical Officer, who expressed the opinion that 50 mg/l as an average concentration is acceptable, provided concentrations remain below 100 mg/l, bearing in mind that the level of 50 mg/l is not an absolute guarantee that infantile methaemoglobinaemia can be avoided, which is the major risk involved.

Another area of controversy with the EC Commission has concerned permission of derogations from the EU drinking water standard. Initially, the government issued about 50 derogations for nitrate from groundwater supplies, on the basis of the nature and structure of the ground, which the Directive allows. The Commission challenged these derogations: in 1992 the European Court of Justice held that the UK had failed to ensure that water supplies in 28 water supply zones in England complied with the standard for nitrate. Since then, programmes to ensure compliance with the standard have been established. Annual monitoring data for 1995 confirmed that these water supply zones now comply with the standard for nitrate. The government also accepted that 50 mg/l should be the maximum acceptable concentration, meaning that each sample must comply with the limit.

During the time period when these issues were debated, the British water supply companies were being privatized: this was an issue since extra costs involved in meeting stricter standards would have to be borne by the water companies. Eventually the EU standards were accepted and became part of the Water Act of 1989. The dispute with the Commission has a clear influence on the setting of these standards.

The main concern in the UK about pollution from agricultural wastes has concerned farm pollution incidents (Table 8.8) such as spills, treatment system failures etc., but not with diffuse, non-point pollution from farms.

The INSC and PARCOM decisions to aim for a 50% reduction of emissions of nitrogen and phosphorus in the period between 1985 and 1995, has initially been accepted by the UK, but this decision has later been reversed. The UK has

Table 8.8 Farm pollution incidents in England and Wales, 1985–1989

Source of pollution	1985	1986	1987	1988	1989
Cows					
slurry stores	717	695	705	801	589
solids stores	185	143	148	194	121
yard/parlour washings	610	816	821	836	578
land run-off	180	244	212	345	380
treatment system failure	116	177	84	96	65
silage liquor	1 006	592	1 003	815	245
Pigs					
slurry stores	164	169	217	231	169
yard washings	85	89	54	59	64
land run-off	5 757	69	74	89	92
treatment systems failure	7	21	21	20	19
Others	383	412	551	655	567
Total	3 510	3 427	3 890	4 141	2 889

Source: NRA/MAFF, 1990

thus accepted the Nitrate Directive standards for drinking water, after some controversy and has accepted the PARCOM Recommendations only partially. As a result of the application of the Nitrate Directive in the UK, 69 vulnerable zones have been designated, covering about 5% of agricultural land. In addition, 32 nitrate sensitive areas have been designated (within the vulnerable zones) under the EC Agri-environment Regulation in which farmers voluntarily make substantial changes in their farming systems to reduce nitrate leaching in return for compensating payments.

The MAFF Code of Good Agricultural Practice for the Protection of Water was published in 1991. The code is an advisory document, but contains advice on slurry storage where there is a legal requirement to have a minimum of 4 months' storage for livestock slurry (including foul yard water) in England and Wales, unless the farmer can show that more frequent land application will not cause water pollution. In Scotland there is a separate Code and the minimum requirement under the Scottish legislation is for 6 months' storage. The MAFF Code advises that livestock manures and other organic wastes should not be spread within 10 metres of a water course, ditch or stream and not within 50 metres of a spring, well or borehole that supplies water for human consumption or is to be used in farm dairies.

The UK has put a lot of emphasis on research and advisory services, which have stressed lower inorganic fertilizer application rates and the efficient utilization of nutrients in manure.

POLICY FORMULATION AND LEGAL INCORPORATION

The Control of Pollution (Silage, Slurry and Agricultural Fuel Oil) Regulations 1991

This secondary legislation was made under the Water Act 1989 and will be amended in early 1997. The Regulations lay down standards for structures used in the making of silage and for the storing of slurry (including foul yard water) and agricultural fuel oil. The amendments to be introduced in 1997 will extend permission for the making of silage in field heaps to sites notified in advance to the Environment Agency where there is no risk of causing water pollution. The making of field silage is not allowed in Scotland.

The Water Resources Act, 1991 (replacing the Water Act, 1989)

Under this law, it is an offence to cause or knowingly permit a discharge of poisonous, noxious or polluting matter or solid matter to any 'controlled water', meaning groundwater, inland fresh water and coastal water. Under the Water Act, it has been possible to prosecute the perpetrators of the worst farm pollution accidents. The maximum fine that can be imposed for water pollution offences in the magistrates' court (the lower court) has been increased from £2 000 to £20 000.

The Environment Protection Act, 1990
Under the Environment Protection Act, nitrate sensitive areas were set up in groundwater catchment areas. Initially, only 10 pilot projects were organized, where farmers were offered payments in return for 5-year agreements to modify farming practices, or to change land use to reduce nitrate leaching, on a voluntary basis. Participation was good, averaging 87%. Later, another 22 areas covering 28 sources entered the scheme. Under this law, the maximum fine that can be imposed for water pollution offences in the magistrates' court (the lower court) has been increased from £2 000 to £20 000.

The Water Supply (Water Quality) Regulations 1989 (as amended)
All suppliers of drinking water are obliged to meet drinking water standards contained in Council Directive 80/778/EEC relating to the quality of water intended for human consumption. In the case of nitrate, all suppliers were asked by the Department of the Environment (DoE) to prepare programmes ensuring that all supplies meet the standard by 1995.

The EC Nitrate Directive
The designated nitrate vulnerable zones, which are planned to become operable by the end of 1999, will include the present NSAs. Nutrient pollution measures that are proposed for these areas will be compulsory and enforceable, but there will be no payments to farmers. However, there is grant aid available for the construction of farm waste (slurry and manure) storage and handling facilities.

The Environment Act, 1995
This primary legislation set up two unitary pollution control authorities in three countries of the United Kingdom: the Environment Agency in England and Wales, and the Scottish Environment Protection Agency in Scotland. These have the powers to control pollution to any or all three of the environmental media, water, air and soil. The Environment Agency was formed by amalgamating the former National Rivers Authority (NRA), Her Majesty's Inspectorate of Pollution (HMIP) and certain waste disposal responsibilities of Local Authorities.

REGULATORY CONTROLS (GENERAL, SPECIFIC, REGIONAL, TANGENTIAL, SELF, CODE OF PRACTICE)

Presently, the only mandatory measures that British farmers have to deal with are minimum requirements for manure and silage storage facilities: these should have non-permeable floors and a means of effluent collection and slurry stores should have a minimum storage capacity for manure of 4 months, unless the farmer can show that more frequent spreading of slurry to land will not cause water pollution. The regulations do not include specific maintenance requirements, but do require that specified performance standards must be met at all times for at least 20 years. Silage making in a field will be allowed to continue in England and Wales subject to a site notification procedure to the Environment Agency, but is not allowed in Scotland.

In the voluntary nitrate sensitive areas, there is a 'basic scheme', with certain requirements relating to low nitrate farming, a 'premium grass scheme' covering the extensification of intensive grassland and a more far-reaching 'premium arable scheme', which involves converting arable land to extensively managed grassland. Under the basic scheme, farmers must:

- Limit nitrogen inputs (inorganic or organic) to 150 kg/ha per annum, or the economic optimum where this is less: in a sub-option, nitrogen inputs may be up to 200 kg/ha in one year of the 5-year agreement and 150 kg/ha for the other 4 years of the agreement.
- Refrain from applying any inorganic fertilizer in the autumn or winter and respect other rules on the timing of applications.
- Plant winter crops or green cover crops on bare land in certain specified circumstances.
- Agree not to convert grassland into arable cropping other than in legitimate cases of grass leys forming part of an arable rotation.
- Keep records of application of organic and inorganic nitrogen fertilizer.
- Where organic nitrogen is utilized, it must have been produced on the farm or another farm within the NSA.

MONITORING AND CONTROL, RECORD KEEPING AND VERIFICATION

Since measures so far have not been very stringent, most control measures have been around 'farm pollution incidents', for which extensive records are kept (Table 8.8). Some of the worst offenders have been prosecuted and increasingly large fines can be imposed. In one well publicised case, a fine of £10 000 was imposed plus a requirement to pay £20 000 in legal costs and fish restocking costs estimated at £27 000. The Environment Agency (formerly NRA) is responsible for pollution control and has declared its intention to prosecute forcefully, aiming to recover full costs of the investigation and the costs of restoring watercourses to their natural state after the incident (Baldock and Bennett, 1991).

Record keeping is at present only required in nature sensitive areas, concerning the application of all nitrogen fertilizers and manures, but will in the future be required in the designated nitrate vulnerable zones.

FINANCIAL MEASURES AND INCENTIVES

Since 1991, the construction of new slurry stores or substantial improvements to existing ones have been covered by the new Regulations. Between 1989 and 1995, a grant scheme was generally available for farm waste storage and handling facilities (50% for eligible items, but subsequently reduced to 25% for last year), which provided financial incentives. Currently, grant aid for farm waste storage and handling facilities is only available to farmers with land in designated nitrate vulnerable zones at the rate of 25% of eligible costs.

In nature sensitive areas, farmers are compensated for the nutrient reducing measures they take. The basic scheme pays £65–105/ha and up to £340–590/ha for a 'premium arable scheme'.

EFFECT OF MEASURES

The nutrient reduction efforts that have been made by the British government have emphasized research and advising farmers to reduce mineral fertilizer applications on a voluntary basis where excessive rates were being applied, and to make better use of manure. Improvement to storage facilities was mandatory and was encouraged by financial incentives. This has led to a substantial reduction in serious point-source water pollution incidents. PARCOM data show that the application rates of manure did not change greatly between 1985 and 1992: total-N went from 74 kg N/ha to 72 kg N/ha. Total-P was 17 kg/ha in 1985 and 18 kg/ha in 1992 (see Table 3.13). The application rates of mineral fertilizer showed a 14% reduction for total-N between 1985 and 1993 and for total-P a reduction of 17% (see Table 3.14). The nitrogen surplus showed an 11% reduction for crops and grassland between 1985 and 1993, and a 16% reduction for all agricultural land. The total-P surplus for crops and grassland has been reduced by 15%, while for all agricultural land it was reduced 17% (see Table 3.15). For total inputs of nutrients from agriculture and expected results of the measures taken or planned, no information is provided, since the UK is not subject to the 50% reduction.

The theoretical effect of measures A1–A5 to reduce run-off on the quality of surface water are moderate in the middle parts of England and in Cornwall (less than 40 mg N/l) and in Ireland. For the rest of the country the improvement will be less, somewhere between 0 and 40 mg N/l (Fig. 8.3; de Cooman *et al.*, 1995). The theoretical effect of measure A6 on the quality of surface water, the improvement will be considerable in the eastern part of the country, due to a reduction in mineral fertilizer and equilibrium fertilization (Fig. 8.4; de Cooman *et al.*, 1995). The theoretical effect on shallow and deep groundwater are to be expected in the same eastern parts of the country, where, according to Figs. 8.5 and 8.6, the improvement will be moderate (40 mg N/l) to considerable (somewhere between 40 and 100 mg N/l).

ROLE OF DIFFERENT ACTORS

The Department of the Environment is responsible for water quality policies. Where policies impact upon agriculture (e.g. implementation of the Nitrate Directive), it shares this responsibility with the Ministry for Agriculture, Fisheries

Figure 8.3 Theoretical effect of measures A1–A5 on the quality of surface water. Source: de Cooman *et al.*, 1995

and Food (MAFF). The Environment Agency grants discharge permits and monitors these permits. All of these organizations are involved with policies concerning eutrophication of the North Sea because the issues involved are complex and reduction measures could potentially be very costly.

The British environmental organizations can be subdivided into two different groups: on one side, there are organizations for nature protection, whose objective it is to protect nature areas, endangered species and country estates. The people active in these organizations are often very influential and have connections with policy makers. On the other hand, there is the Green Movement, including Green Peace and Friends of the Earth, who have fewer connections in government circles. They try to influence policy through the media, public opinion and through parliament.

The agricultural lobby, together with agro-business and fertilizer companies, are well organized and carry a lot of clout; the chemical- and fertilizer companies play a particularly important role. There is a Green Party, which did surprisingly well during elections for the European parliament.

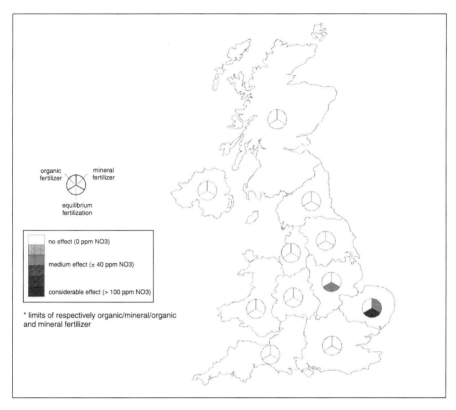

Figure 8.4 Theoretical effect of measure A6 on the quality of surface water. Source: de Cooman *et al.*, 1995

FUTURE DEVELOPMENTS

As required by the EC Nitrate Directive, an Action Programme for the further reduction of nutrient pollution is being finalised, in which the designation of nitrate vulnerable zones is central. For these NVZs a number of mandatory measures have been proposed that deal with:

- Inorganic fertilizer:
  Inorganic fertilizer not to be applied between 1 September and 1 February, unless there is a specific crop requirement.
  No application when soil is frozen hard, waterlogged, flooded or covered with snow.
  No application on steep slopes.
  Not to exceed crop requirement for nitrogen fertilizer, taking account of crop uptake and soil supply from organic matter, crop residues and organic manure.

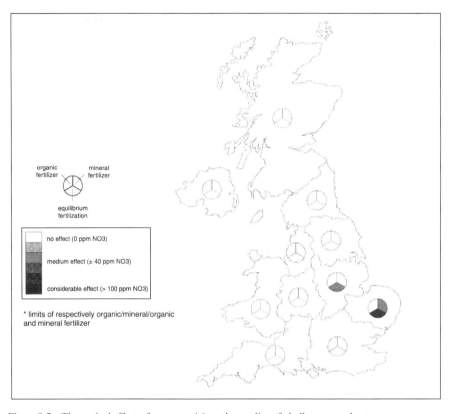

Figure 8.5 Theoretical effect of measure A6 on the quality of shallow groundwater

Not to apply fertilizer in such a way that it will enter directly into surface water.

- Organic manure:

  Application of organic manure shall not exceed 210 kg N/ha averaged over the area of the farm not in grass each year. For the area of grass on the farm, the proposed limit is 250 kg N/ha.

  On sandy or shallow soils, do not apply slurry, poultry manure or sludge to grass fields between 1 September and 1 November, and to fields not in grass between 1 August and 1 November.

  Other application measures are the same as for inorganic fertilizer, with the addition of the requirement not to apply manures within 10 metres of surface water.

  Requirement for all new, substantially reconstructed or substantially enlarged stores for slurry and silage to comply with minimum requirements under the 1991 Regulations.

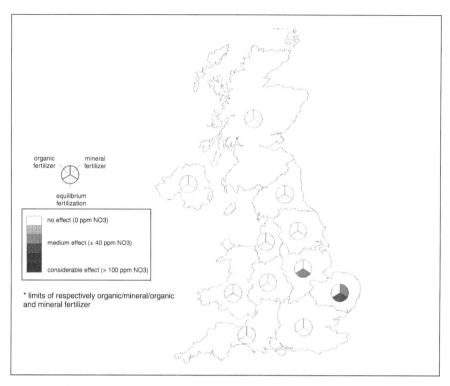

Figure 8.6   Theoretical effect of measure A6 on the quality of deep groundwater. Source: de Cooman *et al.*, 1995

All farms must keep fertilizer plans, with field by field records of application of both organic and inorganic manure.

These measures are planned to go into effect in 1999. The nitrate sensitive areas that are currently operable on a voluntary basis, are included in the area of the designated nitrate vulnerable zones. The NVZs will cover about 5% of the total agricultural area in England and Wales.

UNITED STATES

CHAPTER 9: UNITED STATES

ENVIRONMENTAL PRESSURES AND SOURCES

The amount of nitrogen fixed annually by industrial nitrogen fixation processes and cultivation of nitrogen fixing legumes now exceeds by 10% the amount fixed by terrestrial ecosystems before the advent of agriculture (USEPA, 1993a), reflecting the over-fertilization of the earth. More than 90% of the nitrogen present in the soil is in the organic form and has the -3 oxidation state, which is not mobile. Only a small portion is in the nitrate form with a +5 oxidation state. This form is taken up for synthesis, leaches rapidly and can result in extensive groundwater contamination.

The quantity of organic waste produced annually in the US exceeds 400 million ton/year. The amount of animal manure equals 900 kg dw/person/year (Pierzynski *et al.*, 1994). The nitrogen content of the different manures ranges from 1.3 to 6.0% with a degree of organic nitrogen mineralization of 25 to 70%. The mineralized form is available for crop uptake and can leach into the groundwater.

Feedlot runoff has ammonia concentrations up to 300 mg/l resulting from the urea hydrolysis. Organic nitrogen concentrations up to 600 mg/l were found (USEPA, 1993a), thus showing its pollution potential. The application of poultry manure for corn fertilization in 17 different areas showed that in half of the cases the recommended rate of 5 ton manure/ha was exceeded, resulting in an overdose of 100 kg N/ha. The recovery of the applied N in most cropping systems generally ranges from 30 to 60%: the remainder can leach into groundwaters. The corn crop nitrogen recovery efficiency for poultry manure decreased from 50% to 34% when the application rate was increased from 107 kg N/ha to 321 kg N/ha. This response indicates greater potential environmental losses at higher application rates (Pierzynski *et al.*, 1994).

AMBIENT ENVIRONMENTAL CONDITIONS

The recommended N application rates for irrigated corn in Nebraska resulted in a 16 mg NO_3-N/l groundwater concentration which could be reduced to 10 mg/l when the N application rate was reduced by 70 kg N/ha (Schepers *et al.*, 1991). A survey in the eastern USA found that streams with 50%, 75% and 90% fertilized agricultural watersheds had mean total nitrogen concentrations of 1.08, 1.82 and 5.04 mg/l. Streams and groundwaters from 268 sites in southeast Nebraska ranged from 0.1 to 233 mg/l total nitrogen, with 37% exceeding the maximum drinking water limit of 10 mg/l and many exceeding health advisory levels for livestock of 40 mg N/l (180 mg NO_3/l). In a study of nitrogen sources entering Chesapeake

Bay, 4% was from animal waste and 34% from excess fertilizers (USEPA, 1993b).

A national survey determined that 2.98 million people (of which 43 500 were infants at risk for methaemoglobinemia) were exposed to more than 10 mg/l nitrate-nitrogen from drinking water in community system wells (USEPA, 1992). In the Metropolitan Water District of Southern California 12% of wells exceeded 10 mg N/l and 4% of the annual supply is lost due to the fact that water quality standards for nitrates and salts are violated. A North Carolina survey of more than 9 000 domestic wells showed that 3.2% exceeded 10 mg N/l. An Illinois survey with 286 groundwater samples had a 19.9% exceedance. A survey in Maryland collected 33 groundwater samples near chicken houses and found a maximum concentration of 57 mg N/l and a median of 9 mg N/l (Canter, 1997).

POLICY FORMULATION AND LEGAL INCORPORATION

The general environmental policy is based on the protection of the natural resources, the assurance of beneficial uses of the resources and the limitation of the polluting sources as regulated by the US Environmental Protection Agency (USEPA).

The impact of manure on the environment is only addressed at the Federal level for animal feeding operations, that are considered as point sources to the receiving surface waters. Extensive livestock grazing is viewed as non-point sources to streams. Excess nitrate concentrations in groundwater are controlled through drinking water regulations of the Safe Drinking Water Act (SDWA).

The Federal Water Pollution Control Act (FWPCA) of 1965 started the legislation for pollution controls based on water quality and beneficial use goals (potable water supply, fish habitats etc.). National technology based standards were established that require secondary treatment levels. Advanced treatment was prescribed for specific water bodies of high quality. The re-enactment of the Federal Clean Water Act in 1987 allowed the States to establish water quality standards and designate water uses. Waste load allocations are made through water quality modeling of water bodies. The Coastal Zone Act Reauthorization Amendments of 1990 (CZARA) regulate feedlots in the coastal zone and non-point agricultural runoff.

The Clean Water Act (CWA) defines concentrated animal feeding operations (CAFO) with more than 1000 animal units (or 301–1000 units with a man-made conveyance) as point sources requiring a discharge permit since 1972. In 1987 the CWA included the regulation of storm water discharges from CAFOs. In 1995 the USEPA developed a guidance manual for more uniform permitting (USEPA, 1995). This Act, as amended, also requires the establishment of non-point source management programs per State and per major watershed in cooperation with the local authorities. Sources that pollute surface waters are required to show best management practices, while taking into account the impact of the practice on groundwater quality.

The application of the Clean Water Act in the dry western USA focuses on control of riparian lands adjacent to creeks, streams and rivers, making up about 1% of the area. These areas have most of the vegetation diversity and about 75% of the wildlife species. They prevent bank erosion losses and are important to control the quality of the river water and the fish habitats. These areas can change greatly by excess grazing, trampling and fertilization from nutrients in manure (Bauer and Burton, 1993). Proposed innovative regulatory approaches to reduce groundwater contamination from excess fertilizer and manure application include an excise tax on fertilizers, limited exchangeable rights to buy nitrogen fertilizer, per acre restrictions on commercial fertilizer application and zoning regulations that limit the type of land use and fertilizer input (Canter, 1997).

REGULATORY CONTROLS

Both ongoing and empty CAFOs with animal feeding operations for at least 45 days per year and no crop growth need a discharge permit when they generate dischargeable waste. These include feedlots, dairy farms, stockyards and auction houses. A point source is a discreet conveyance of pollutants to surface waters through pipes, ditches, tunnels, etc. The regulations define 'waters of the US' as navigable waters, tributaries to interstate waters, wetlands and groundwaters with a direct hydrological connection to surface water. Land spreading of manure from CAFOs is covered under the Clean Water Act and often includes, a nutrient management plan (USDA, 1992).

The discharge permit is issued under the National Pollutant Discharge Elimination System (NPDES) requiring best available technology economically achievable (BAT or BATEA) and best conventional pollution control technology (BCT) based on the best professional judgement (BPJ) of the permit writers. The permit often requires best management practices (BMP), pollution prevention plans and annual reporting of monitoring results. The effluent limitation guidelines for large operations require a retention structure to hold the wastewater and runoff from the 25-year, 24 hour storm event.

Under the initial Coastal Zone Management Act (CZMA) of 1972, 35 States and territories have coastal waters, requiring coastal zone management plans including non-point pollution controls to be approved by EPA and the National Oceanic and Atmopheric Administration (NOAA). The CZARA of 1990 requires runoff controls, waste utilization of the stored waste and a nutrient management plan. The nutrient management plan establishes nutrient input and uptake analysis. It also takes field limitations into account such as sinkholes, shallow soils over fractured bedrock, soils with a high leaching potential, proximity to surface waters, soil erodibility and shallow aquifers (USEPA, 1993c).

The SDWA has the Underground Injection Control Program (UICP). Injection wells accepting feedlot runoff have inventory requirements and performance standards that prohibit movement of pollutants into an aquifer if that causes violation of drinking water standards or adversely affects human health. The Sole Source Aquifer Program (SSAP) requires identification of point and non-point sources of groundwater degradation when areas only have one aquifer. Feedlots located in such an area may need additional management practices. The State Wellhead Protection Areas Program (WPAP) requires States to develop programmes to protect wellhead areas from contaminants that may adversely affect human health, such as pathogens from feedlots infiltrating in such areas. The Surface Water Treatment Rule (SWTR) requires watershed management practices.

Technical N and P reduction measures are generally part of the Best Management Practices formulated by the different states, and several studies have documented the effectiveness of these approaches. A study showed that runoff losses of N from bare soil with a 5% slope were reduced by 67% to 13 g N/h per meter of width, with the use of a 2.7 meter strip containing residue corn stalk cover to trap sediments and nutrients. A 1.80 wide strip reduced the incoming nitrogen by 20% (Alberts *et al.*, 1981). Runoff losses of fertilizer P from contoured corn in Minnesota were 1.3% at application rates of 66 kg P/ha/year. No-till corn reduced P losses by 23–27% compared with tilled controls in Mississippi (Sharpley *et al.*, 1995).

STATE REGULATIONS FOR MANURE CONTROLS

As part of the current study, questionnaires were sent to each of the 50 states of the USA to determine whether they had manure control programmes. The outcome is listed in Table 9.1. The result of the Canadian Province of British Columbia is also included. Twelve states had regulations that limited the manure application to the land. Only four states require a manure bookkeeping. One State had regulatory or financial incentives for installation of advanced technical equipment and measures. The states with programmes enacted 56 manure control approaches (38% of all possible measures). These are discussed per state. An overview of the agricultural non-point source pollution management programmes in 15 states is given in Table 9.2. It includes measures for erosion control, water conservation and animal waste systems.

Alabama
Alabama has a statewide policy for manure/mineral management that advocates a long term equilibrium between application and uptake. There are regulations that limit the application of manure to the land. No manure limitations have yet been formulated for wellhead protection areas.

Alaska
Alaska has no regulations for manure as only 1700 sheep, 10 200 cattle and 2000 hogs reside in the state. It does advocate the use of animal manure as a fertilizer, using nutrient values of Tables 9.3 and 9.4. Moose droppings in the summer contain 2.5% N, 1.84% P_2O_5 and 1.2% K_2O. It is recommended that spreading on frozen ground and within 30 meters (100 feet) of streams, lakes or ponds should

Table 9.1 Mineral/Manure policies in the U.S. and Canada (State and Provincial policies) Questionnaire results:

Measures	Alabama	Alaska	Arizona	Illinois	Kansas	Kentucky	Minnesota	Mississippi	New Jersey	New Hampshire	North Dakota	Ohio	Canl British Columbia	Row Totals Yes
A. Technical Source Reduction Measures														
1. Are there regulations to limit application to land?	Yes (1)	No (3)	Yes	Yes	Yes	Yes	Yes	Yes	Yes	Yes	Yes	Yes	Yes	12
2. Is mineral/manure application accounting obligatory?	No	No	No	No	Yes	No	Yes (4)	No	No	Yes (8)	No	Yes	No	4
3. Regulations/financial incentives to reduce ammonia emissions	No	No	No	No	No	No	No	No	No	No	No	Yes	No	1
B. Regulatory Measures														
4. Taxing or fee system for excess manure production	No	No	No	No	No	No	No	No	No	No	No	No	No	0
5. Ground water protection measures limiting manure/minerals application	No	No	Yes	Yes	Yes	Yes	Yes	Yes	Yes	Yes	No	Yes	Yes	10
6. Eutrophication reduction measures limiting manure/minerals	No	No	No	No	Yes	Yes	No	No	No	No	Yes (9)	Yes	Yes	5
7. Regulations for special area's (watersheds, ecotopes, P-saturated soils, leachable sandy soils) limiting manure/minerals application?	No (2)	No	No	No	No	Yes	No (5)	No	Yes	No	No	No	No	2
8. If yes, based on manure volume (v), NPK balance (NPK), or both (b)?	N/A	N/A	N/A	N/A	B	V	N/A	NPK	–	N/A	NPK	B	NPK	NPK 3x B 2x V 1x
C. Policy formulation														
9. Is there a statewide policy for manure/minerals management?	Yes	No	No	No	Yes	Yes	Yes (6)	Yes	Yes	Yes	No	Yes	Yes	8
10. Is there legal incorporation of policy in Statutes?	No	No	No	No	Yes	No	No	Yes	No	Yes	No	No	Yes	5
11. Does policy advocate long term equilibrium manure/fertilizer applic., where input=crop uptake plus loss term?	Yes	No	No	No	Yes	No	Yes (7)	Yes	Yes	Yes	No	Yes	Yes	8
Total number of Yes answers per State	3	0	2	2	7	5	5	5	6	6	2	7	6	56

1. Technical criteria only
2. not at present – potential for development in wellhead area's
3. Total 1995 livestock population in Alaska: 10200 cattle, 1700 sheep, 2000 hogs
4. Typically 3 year records for permitted facilities.
5. Guidelines for land near surface water
6. A nitrogen management plan
7. Nitrogen based rates
8. Only if they receive federal funds for practices.
9. N.D. has regulations that specify that livestocks waste shall not be placed in waters of the state or be put in a place where they could reach waters of the state

Table 9.2 Agricultural non-point source pollution management. Many states have agricultural nonpoint source reduction programs. Most of them are cost share programs funded by the General Fund. Here are brief descriptions of selected programs that show a variety of operational, administrative and funding element.

Alabama	Cost sharing for soil and water conservation, agricultural water quality and reforestation. Administered by Agricultural and Conservation Development Commission through the State Soil and Water Conservation Committee. Funded by general fund.
Arkansas	Tax credit, up to $3,000 per year, for construction of ponds, lakes or other water control structure used for irrigation, water supply, sediment control, agriculture or water management. Administered by Soil and Water Conservation Commission. Funded by State Income Tax Credit.
Connecticut	Cost sharing for animal waste systems. Administered by Department of Agriculture through Agricultural Stabilization and Conservation Service. Funded by General Assembly annual appropriation. Cost sharing and regulation program for specific aquifers. Farmers must follow resource management plans. Administered by Department of Environmental Protection through Soil and Water Conservation Districts. Funded by CWA 319 grants.
Delaware	Cost sharing for erosion and sediment control, water quality, organic waste systems, water management, forestry, wildlife habitat development and others. Administered by Department of Natural Resources and Enviroment Control. Funded by Bond Act.
Florida	Cost sharing for dairy operations in the Lower Kissimmee River Basin for animal waste management systems. Administered by Department of Agriculture and Consumer Services, Bureau of Soil and Water Conservation. Funded by General Fund.
Idaho	Cost sharing for farms in approved project areas, identified by State Agricultural Water Quality Plan. Administered by Department of Health and Welfare and Soil Conservation Commission through Soil Conservation Districts. Funded by Water Pollution Control Fund financed by taxes on cigarettes, alcohol, inheritance and sales tax. Long-term, low-interest loans to farmers and ranchers for conservation improvements. Administered by Soil Conservation Commission through Soil Conservation Districts. Funded by inheritance taxes.
Illinois	Cost sharing for farmers who have had a complaint lodged against them under the Illinois Erosion Control Law. Administered by Department of Agriculture, Division of Natural Resources. Funded by General Fund. Cost sharing for constructing enduring practices under the County Conservation Practices Program and the Watershed Land Treatment Program for soil erosion control. Administered by Department of Agriculture, Division of Natural Resources. Funded by General Fund.
Iowa	Cost sharing for erosion control pratices with special provisions for one time payments for minimum tillage and stripcropping and higher rates for special watersheds. Administered by Department of Agriculture and Land Stewardship, Division of Soil Conservation Districts. Funded by General Fund.
Kansas	Up to 70% cost sharing for water conservation practices to improve water quality and quantity by the reduction of soil, water and nurtients loss from the land and for implementation of the 1985 Food Security Act. Administered by the State Conservation Commission through Conservation Districts. Funded by General Fund and Dedicated Water Plan Fund (a fund supported by fees on water supplied by public water supply systems).

Table 9.2 *Continued*

Maryland	Up to 87.5% cost share for approved best management practices installed for agricultural pollution control. Participant can receive both federal ACP funding and state funding up to maximum set by state. Administered by Department of Agriculture and Department of the Environment through Conservation Districts. Funded by Chesapeake Bay Water Quality Loan Act of 1988.
	State Conservation Reserve Program – provides addditional $20 per acre for land enrolled the federal CRP if it is either in the Chesapeake Bay Area, or is a vegetative filter strip adjacent to a stream. Allows a one-time $100 per acre bonus if planted to trees. Administered by the Department of Agriculture. Funded by the General Fund.
Minnesota	Property tax credit program – farmers located in exclusive agriculture zones can receive a property tax rebate of $1.50 per acre per year. Administered by local county. Funded by conservation fund derived from $5.00 surcharge on mortage and deed recordings.
Missouri	Up to 75% cost share for eligible practices in conservation plan. Cost share for lands eroding above tolerable soil loss limits, plus other special areas to encourage less intensive land uses. Administered by Soil and Water Districts Commission. Funded by 0.1% sales tax for soil and water conservation.
	Rebate on interest costs of conservation loans acquired from lending institutions. Administered by Department of Natural Resources, Soil and Water District Commission. Funded by interest drawn on state fund investments.
	Special Area Land Treatment (SALT)- high priority watershed program combining two programs described above. Administered by Soil and Water Districts Commission through Conservation Districts. Funded by 0.1% sales tax and interest on state fund investments. Projects are funded for 5 years. Districts receive $7 000 per year for personnel, equipment, tours and newsletters.
Montana	Loans up to $35 000 for any conservation practice. Can use loan to match other cost share programs. Administered by Rosebud Conservation District. Funded by 1.5 mill levy on real property.
N. Carolina	Provides a state income tax credit of 25% up to $2 500 per year for the purchase of conservation tillage equipment. The tax credit may not exceed the tax liability for the year. Excessive credit can be carried forward for next 5 tax years. Administered by Tax Commission.
	Tar-Pamlico watershed – point source facility operators are allowed to achieve nutrient loading limits by financing BMP installation on farms in the watershed.
Utah	Revolving loan fund – provides low interest loans (3% interest with 4% administrative fee) for agriculture and energy conservation, range improvement and watershed development. Administered by Soil Conservation Commission. Started in 1983. Funding from General Fund of $747, 187 in FY87–88. $1 726 686 loan repayment and interest in FY87–88. Current balance in fund of $14 450 000.

Table 9.3 Animal manure as fertilizer (Figures given in pounds/ton. Includes solid, liquid, and bedding.)

Animal	Percent moisture	Tons of manure produced/year per 1000wt** (Fresh: normal bedding)	Approximate composition pounds per ton			Percent Organic Matter
			N	P_2O_5	K_2O	
Cow	86	15.0	11	3	10	30
Goat	65	–	16	10	17	60
Duck	61	–	22	29	10	50
Goose	67	4.5	22	11	10	50
Hen	73	–	22	18	10	50
Hog	87	18.0	11	6	9	60
Horse	80	10.0	13	5	10	60
Sheep	68	7.5	20	15	8	60
Steer	75	8.5	12	7	11	60
Turkey	74	–	26	14	10	50
Rabbit (dried)	6	6.0	45	27	16	–

* Western Fertilizer Handbook, California Fertilizer Association, 6th Edition
** Based on animals confined to stalls year-around.
Source: Purser, 1994

be avoided. The applied amount should be limited to 12 tons/acre or 550 pounds/1000 square feet.

Arizona

The Arizona Environmental Act of 1986 established the Department of Environmental Quality. This is charged with developing a program of Best Management Practices (BMP) for Regulated Agricultural Activities (RAA) through a general rather than individual permit programme. It sets up an enforcement framework for user classes, cooperation with the regulated community and voluntary compliance. Guidance material for BMP was developed for nitrogen fertilizer management in cotton, rangeland and vegetable crops, together with an interactive nitrogen management computer programme. Research established 7 BMPs and 34 specific guidance practices to limit N losses from irrigated croplands and 22 individual crop guidances.

Joint agencies have established a nitrogen monitoring programme for ground- and surface waters of the State. The aquifer protection programme requires permits for discharge to an aquifer from impoundments, pits, ponds, injection facilities, leaching and tailing operations, and land treatment facilities to protect the groundwater table, which is generally below 250 feet. The facilities have to employ best available demonstrated control technology (BADCT) while not violating aquifer water quality standards.

Table 9.4 Micro-nutrients in Animal Manure (pounds/ton)

Animal	Boron	Calcium	Copper	Iron	Magnesium	Manganese	Molybdenium	Sulphur	Zinc
Horses	0.03	15.7	0.01	0.27	2.8	0.02	0.002	1.4	0.03
Cattle	0.03	5.6	0.01	0.08	2.2	0.02	0.002	1.0	0.03
Sheep	0.02	11.7	0.01	0.32	3.7	0.02	0.002	1.0	0.05
Hogs	0.08	11.4	0.01	5.6	1.6	0.04	0.002	2.7	0.12
Laying Hens	0.12	74.0	0.03	0.93	5.8	0.18	0.011	6.2	0.18
Broilers	0.08	29.0	0.06	2.00	8.4	0.46	0.007	–	0.25

Source: USDA Research Data, Ohio

Delaware

Delaware has a non-regulatory process for animal agriculture, which does not require permits except for CAFO's according to federal regulations. The State has guidance manuals for manure management for livestock farmers. Delaware has a provision in the Water and Air Resources Act as amended in 1993 to provide compensation for a contaminated drinking water supply. If a state or federal drinking water standard is exceeded for any contaminant, except for bacteria, viruses, nitrates and pesticides when applied according to manufacturers instructions, then the likely source has to provide remedial actions or install an alternative drinking water supply at his cost.

Illinois

Illinois has regulations to limit manure applications to the land and groundwater protection measures.

Kansas

Kansas has extensive regulations to control release of animal waste and runoff from confined livestock feeding operations. The permitted site should be located as far as practical from residences, streams and water supplies and not be in a 10 year flood plain. A minimum separation distance of 30 meters (100 feet) is required between the livestock feeding operation and the perimeter fence, water supply wells and reservoirs. The confined livestock feeding facility for less than 100 cows should be located more than 1320 feet or 1/4 mile (450 m) from the nearest habitation, which increases to 5280 feet for more than 500 head. The lowest point of the feedlot should be 10 feet above groundwater aquifers or seasonal perched water table and the lot should have less than 5% slope.

The State requires a minimum of 40 acres (16.2 ha) for livestock operations with one acre of land per acre of lot for the disposal of stormwater runoff, or 3 acres of land per acre-feet of runoff to control the 25-year 24-hour storm. The land disposal area must be 1 acre per 100 head of livestock or 1 acre per 20 ton d.w. per year of livestock wastes. The application rates should be based on the nutrient requirements of both the land and crops. The runoff disposal system should be able to handle 1/10 of the runoff storage volume each day of operation. The thereby supplied nitrogen should be restricted to 250 lbs N/acre as required for crop production.

Liquid manure facilities should have minimum storage capacity for 120 days of dry weatherflow, with additional storage for the 25-year 24-hour storm. The most common waste water treatment facilities are anaerobic lagoons or algal ponds.

Kentucky

More than half of Kentucky's rivers are partially or wholly closed to swimming due to degraded water quality. Agriculture contributes 27% to the non-point source pollution and impaired 25% of the lake acreage use. New animal waste management systems were required for 51% of the 4467 livestock operations.

The Commonwealth of Kentucky enacted the comprehensive Agriculture Water Quality Act (Senate Bill 241) in 1994 to control non-point sources from agriculture, silviculture and turfgrass industry and to protect surface- and groundwater. Senate Bill 377 established a state funded cost share program in 1994 for control measures and provides 50–75% cost share for sedimentation and pollution control and wildlife conservation. The WQA focuses on pollution prevention and has enforceable best management practices which are technology driven. The 'bad actor clause' and the 'corrective measure plan' are essential tools. The Agriculture Water Quality Authority developed best management practices.

The AWQA applies to all farms larger than 10 acres, requiring a water quality plan for each. The Best Management Practice regulates the waste utilization on land, including poultry litter. The waste should not be applied to frozen or snow covered soil unless there are no reasonable alternatives. The amount of applied waste should equal the N, P, K requirement of the crop and not exceed the soil capacity to assimilate and decompose the waste. The application of liquid waste should be limited to 1/2 inch per hour when irrigated and not exceed the field capacity at the time of application and not be applied 48 hours after rain or within 12 hours of forecasted rain.

A filter strip of 30 feet should be maintained around any animal waste application area adjacent to streams, ponds or lakes. Vegetation near streams should not be degraded by cattle grazing to maintain soil erosion control and filtering and uptake benefits.

Minnesota

The livestock producers generate $3.8 billion in farm income and the same amount in the value added in downstream industries with 63 000 direct jobs and 93 000 jobs in the related industries. Wells are used by 70% of the population for drinking water supply and are vulnerable for livestock contamination.

The Minnesota Pollution Control Agency (MCPA) is the permitting body for feedlots that are new or remodeled with 50 animal units or more. These feedlots also have manure management plans as part of their permit. County Boards can assume the responsibility for permitting. Location planning and siting were partially shifted from the state level to the counties because of their landuse planning activities. An environmental assessment is required for more than 2000 animal units.

The Minnesota Department of Agriculture (MDA) has developed the guidelines for the Best Management Practice. Land application rates of manure shall not exceed local agricultural crop nutrient requirements except where allowed by permit. The guidelines are voluntary but have been incorporated in some local feedlot ordinances and individual permits. The manure application is limited by its relative most abundant component. As an example the MDA calculates that

Table 9.5 Recommended Separation Distance in Feet

	Surface spreading	Incorporation or injection	Irrigation
Streams or rivers	*	50	200
Lakes	*	100	300
Water well	200	200	200
Sinkholes	100	50	200
Individual dwellings**	100	50	300
Residential development	300	300	1000
Public roadways	25	10	300

* See Table 9.6

** Distance may be reduced with owners permission.

MPCA recommended 1995

Source: Reichow, 1995

a 60 cow dairy should provide enough fertilization for 200 acres of cropland at a fertilizer saving of $10 245 per year to give a 135 bushel yield of a rotation crop of oats, 2 alfalfa and 2 corn.

The feedlot is not allowed in a floodplain with a 100 year recurrence or shoreland (within 1000 feet from a lake or 300 feet from a river) or in the drainage area of a sinkhole or shallow soils overlying fractured bedrock. The set-back distances for land application of manure are given in Tables 9.5 and 9.6. Special attention is given to the 1.6 million abandoned wells that have punctured protective layers and are often not sealed and plugged, allowing livestock waste to infiltrate into aquifers.

Table 9.6 Separation Distance from Surface Waters for Surface Application

Slope (%)	Soil Texture	Time of Year	Minimum Separation
0–6	coarse	May–October	100 feet
0–6	coarse	November–April	200 feet
0–6	medium-fine	May–October	200 feet
0–6	medium-fine	November–April	300 feet
>6	coarse	May–October	200 feet
>6	medium-fine	May–October	300 feet
>6	all soils	November–April	not recommended

MPCA recommendations 1995

Source: Reichow, 1995

Mississippi

The Mississippi Soil and Plant Amendment Law of 1978 regulates the application of animal manure as a plant amendment. The State law regulating NPDES waste water discharge permits also requires that the feedlot operation or manure treatment plant is not closer than 1000 feet from the nearest dwelling and 300 feet from the nearest property line. The distances for broiler operations are 600 and 150 feet respectively, and for land application of manure 300 and 50 feet respectively.

New Jersey

The State of New Jersey regulates discharges from feedlots through a permitting system. The land application of manure is regulated by the Department of Agriculture on the basis of a state wide policy advocating equilibrium between input and output. The land application also has to comply with regulations for coastal zone management, local land use and solid waste control.

New Hampshire

The Department of Agriculture of the State of New Hampshire recommends that accurate recording systems be developed for crop yield and manure application rates as part of its best management practices. The maximum benefit for the manure is achieved by proper calibration of manure equipment and incorporation into the soil after application, which reduces losses, bacterial contamination and water quality impairment.

An example calculation by the DASNH shows that a typical fertilizer program for silage corn includes a pre-plant incorporation of 15–20 tons of dairy manure per acre, the use of a starter fertilizer at planting and the application of supplemental nitrogen at the 8–16 inch stage of plant growth based on crop need as determined by soil nitrate testing.

New Mexico

New Mexico does not have regulations to limit manure applications to the land but they do have extensive regulations to limit ground water discharge of milking barn washing operations because of high nitrogen content.

North Dakota

The state has regulations to limit the manure application to the land and to prevent eutrophication. Livestock waste should not be placed in a location where it could reach surface water.

Ohio

Ohio has 56 000 farms, 1.6 million head of cattle, 0.3 million dairy cows, 1.8 million hogs, 23 million chicken and 5 million turkeys. The livestock industry generates 80 000 tons of P_2O_5 equal to 30% of the crop requirement in the state.

The ammonia from livestock waste is toxic to aquatic life and can destroy entire fish populations. In 1989 16 fish kills from manure discharge resulted in

Table 9.7 Recommended maximum manure-application rates at different soil test levels[a]

Bray P1 Level (lb P/A)	Surface applied on high-runoff potential sites [b]	Incorporated or low-runoff potential sites [c]
0 – 60	N needs of non-legume crops	N needs of non-legume crops
	N removal rate of legume crops	N removal rate of legume crops
60 – 250 [d]	N needs or P removal rate for non-legume crops, whichever is less	N needs of non-legume crops
	N or P removal rate for legume crops, whichever is less	N removal rate for legume crops
250 – 300 [d]	Manure application for crop production purposes not recommended	N needs or P removal rate for non-legume crops, whichever is less
	N or P removal rate for legume crops, whichever is less	
>300 [d]	Manure application for crop production purposes not recommended	Manure application for crop production purposes is not recommended. If application is necessary, apply no more manure than supplies N or P removal for the next crop, whichever is less. A site plan that controls erosion and runoff is recommended

a Application of manure at rates above these recommendations may require approval and/or permits by appropriate government agenices.

b Surface application is any application at a depth that would be disturbed by tillage within the nest three years. High-runoff potential refers to sites where surface movement of manure and/or phosphorus is likely to occur from the field of application.

c Incorporation is any application incorporated at a depth that would not be disturbed by tillage within the next three years.
   Low-runoff potential refers to sites where surface movement of manure and/or phosphorus from the field of application is not likely to occur under normal weather conditions

d Yearly plant tissue and soil analysis recommended.

20 000 fish deaths. The N and P from manure cause excessive algal growth in streams and ponds. Livestock waste is affecting 4132 stream miles. More than 200 environmental complaints are registered annually.

Farms with more than 1000 animal units (equivalent to feeder cattle) are regulated by the Ohio EPA under the Ohio Water Pollution Control Law and need a waste management plan, a permit to install animal waste treatment facilities, a discharge permit when discharging to waters of the state and a storm water permit with more than 5 acres generating runoff. Failure to comply results in civil penalties of $10 000 per day or criminal penalties of $25 000 per day and one year in prison. For smaller farms the storage, handling and manure application is regulated by the Ohio Department of Natural Resources. Common livestock manure treatment units include sedimentation, filtration, drying, coagulation, anaerobic lagoons, aerobic lagoons and composting.

The Best Management Practices are included in the Waste Management Guide

Table 9.8 Approximate amounts of plant nutrients removed in harvested crop

Crop	Nutrients removed for given yield in lb/acre		
	N	P_2O_5	K_2O
Alfalfa (6 T)	340*	80	360
Corn (150 Bu)			
Grain	135	55	40
Stover	100	25	160
Corn-silage (26 T)	235	80	235
Grass-cool season (3.5 T)	140	45	175
Oats (100 Bu)			
Grain	65	25	20
Straw	35	15	100
Sorghum-grain (7 600 lb)			
Grain	105	30	30
Stover	80	50	230
Soybean (50 Bu)	190*	40	70
Sugar beets-roots (25 T)	100	50	250
Tobacco-burley and cigar filler			
Leaf (3000 lb)	105	25	185
Stems and suckers (2,000 lb)	55	15	65
Wheat (55 Bu)			
Grain	70	35	20
Straw	30	5	50

* Inoculated legumes fix nitrogen from the air

of the Extension Service and in the Technical Guide of the Soil Conservation Service. Cultural practices of the BMP include application rates, crop rotation, injection practices and regular feedlot scraping. Vegetative practices include grass filter strips, pasture renovation and critical area seeding. Structural practices include storage ponds, stacking facilities, fencing and diversions. A 6–12 month manure storage is recommended, which pays itself back in 2–5 years because of saved nutrients. This also allows the farmer greater flexibility to apply the manure when weather conditions are optimal. Animal feedlots are recommended to have concrete structures and floors to minimize groundwater infiltration of nutrients and bacteria.

The Ohio EPA has a funding program of low interest loans for projects that improve water quality, erosion control, prevention of nutrient and pesticides runoff, stream exclusion fencing etc. Federal funds are available for local watershed non-point source projects. Research has developed treatment processes such as sand filters, wetland treatment systems, composting, pelletising and manure brokerages. An average dairy with 90 heads requires control facilities costing $12 000, while a community of 2000 persons with the same pollution load requires a treatment plant of $2 million. Investments in the former are, therefore, most cost-effective.

The Ohio State Cooperative Extension Service determined that crop yields rarely increased as a result of additional P when soil levels exceeded 60 lbs P/acre (Table 9.7) and that applications above 300 lb p/acre are not recommended (Veen and Huizen *et al.*, 1992). The nutrient removal through harvested crops ranges from 0.09 to 0.64 lbs P_2O_5/bushel (Table 9.8), indicating the need for individual nutrient balance calculations and fertilization schedules. Different agencies use either N or P to determine manure application rates, but it was recommended to use P as that element requires more land and is therefore more protective of the environment.

CANADA

CHAPTER 10: CANADA

ENVIRONMENTAL PRESSURES AND SOURCES

Canada has 12 million cattle, 10.9 million pigs, 0.7 million sheep and 83 million chickens on 272 000 farms in its 10 Provinces and three Territories (Patni, 1994). Most animal operations are located in Quebec and Ontario and are generally confined operations; 20% of the farms contain 80% of the livestock. Nutrient surpluses exist in portions of British Columbia and Quebec. Livestock in Canada generates 235 000 metric tons of nitrogen, 150 000 tons of P_2O_5, and 500 000 tons of K_2O in the manure. Mineral fertilizer consumption is 1 160 000 metric tons of nitrogen, 614 369 tons of P_2O_5, and 356 142 tons of K_2O. Since 1970 the use of nitrogenous fertilizers has increased by 300%. Canada now uses 26.1 kg N/ha, compared with 250 kg N/ha in the Netherlands (Smith, 1992).

Studies showed that 25% of the nutrients in the animal feed are taken up by the animals, while 75% is excreted in the manure. About half of the nitrogen is in the liquid portion, while most of the phosphorus is in the solid fraction, requiring disposal practices for both the liquid and solid portion of the manure. Only 50% of the phosphorus and nitrogen is available for plant uptake when the manure is applied to the soil. Until 1985 the nutrient exports through crop harvesting exceeded soil nutrient additions as shown in Figure 10.1 (CFI, 1990)

AMBIENT ENVIRONMENTAL CONDITIONS

Fertilization of fields increases nitrate levels in ground water. A fertilizer increase from 35 to 200 kg N/ha of corn increased nitrate-nitrogen levels in tile drains from 17 to 55 mg/l in southwestern Ontario (CFI, 1990). Nitrate concentrations exceeding 10 mg N/l were detected in 2.7% of the 183 wells sampled in northeastern Kings County in Nova Scotia (Richards and Milburn, 1992). In 12% of 234 wells sampled in Kings County elevated levels above 10 mg N/l were found. In the entire province 7% of the wells exceeded that value. In the St. Andre region of New Brunswick 39% of the wells exceeded 10 mg NO_3-N/l in agricultural areas.

The nitrate-nitrogen values in groundwaters in Prince Edward Island (Somers, 1992) were highest in row crop areas (5.5 mg/l) followed by feedlot operations with on-site manure storage (5.3 mg/l), as shown in Table 10.1. Nitrate-nitrogen concentrations in excess of 30 mg N/l were found in groundwater below 12 meters in a coarse-textured soil receiving high rates of manure for several years (Miller and Goss, 1992). Annual manure applications of 20–60 tons/ha in Alberta resulted in nitrate-nitrogen values all exceeding 10 mg/l in shallow groundwater (Paterson and Lindwall, 1992). Racz (1992) noted 23–62 mg NO_3-N/l in groundwater beneath

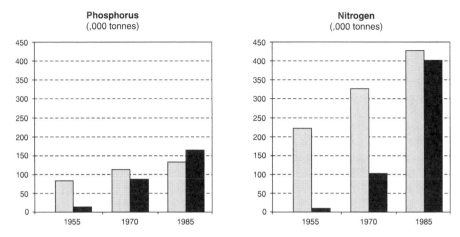

Figure 10.1 Nutrient removals and additions for Alberta

a feedlot on a coarse-textured soil and concentration decreased away from the site due to active denitrification. Wells near farms with on-site manure storage were 5.3 mg N/l, compared with 1.2 mg N/l in non-agricultural areas.

A strong correlation ($R^2 = 0.88$) was observed between total N in the river mouth and animal densities in Quebec river basins. The correlation was 0.77 for phosphorous and suspended solids (Asselin and Madramootoo, 1992). Phosphorus concentrations of as high as 16.2 mg total P/l were measured in runoff from a cattle over-wintering site and 4.6 mg total P/l in a downstream creek (Nagpal, 1992)

POLICY FORMULATION AND LEGAL INCORPORATION

At the Federal level, only the protection of the fish habitat is regulated. The Fisheries Act prohibits the unauthorized discharge of any material that may be harmful to fish into any body of water frequented by fish immediately or at some

Table 10.1 Nitrate concentrations (mg/l NO_3-N) in P.E.I. groundwaters by land use class (monthly sampling results, 54 sites)

Land use categories	mean	max	min	var
All sites	3.98	15.5	<0.2	9.11
Row crop areas	5.57	15.0	2.1	13.74
Non-row crop areas	3.97	10.5	0.7	4.32
Feed-lot operations with on-site manure storage	5.30	13.0	<0.2	6.20
Non-cropped areas (pristine areas)	1.15	5.5	<0.2	1.07
Subdivisions with on-site sewage disposal	4.25	15.5	<0.2	12.40
Subdivisions with central sewage disposal	2.64	6.5	<0.2	2.21

later time. The definition of fish includes parts of fish, shellfish, crustaceans, marine animals, the eggs, sperm, spawn, larvae, spat and juvenile stages of fish, shellfish, crustaceans and marine animals. Punishment of offenders, upon conviction, could include fines up to one million dollars and imprisonment up to 3 years (Patni, 1994).

Control and legislation regarding pollution and nuisance from farm activities is mainly a provincial responsibility. All provinces have legislation that prohibits pollution of water courses.

Common features of the provincial legislation (Patni, 1994) are:

- prohibition of discharge of polluting substances into water bodies;
- authorization for local governments to enact local land use and zoning by-laws;
- protection of public health;
- approval requirement for farm facilities (new, expansion, modification);
- imposition of fines and imprisonment for offenders;
- protection of farmers from nuisance and frivolous litigation ('right to farm').

REGULATORY CONTROLS

Four provinces (Alberta, British Columbia, Nova Scotia and Ontario) have voluntary codes of practice (with regulations in case of non-compliance) while the other provinces and territoria rely on regulations. Producers complying with the code of practice generally do not have to obtain permits for waste discharge and are mostly monitored by their peers. Where regulations or regulatory guidelines apply, the permitting of all producers and enforcement takes place.

The 'Codes of Practice' give guidelines on recommended procedures for siting livestock facilities, for manure management (storage and applications) and setback distances and other technical recommendations. Surface application through irrigation on sloping or crusted soils should not exceed 2000 gallons per acre per application.

PROVINCIAL REGULATIONS FOR MANURE CONTROLS

A brief overview is provided for each of the provinces with significant legislation and controls on manure applications. The overview is made from material obtained from the provinces and from the federal Ministry of Agriculture and Agri-Food Canada (MAAC, 1994).

Alberta
Alberta currently has 4800 cattle feedlots, 3300 of which are located in or adjacent to irrigated areas. The province of Alberta has the Planning Act, the Public Health Act, the Confined Livestock Facilities Waste Management Code of Practice, the Agricultural Operations Practices Act and the Environmental Protection and Enhancement Act (EPEA) of 1993 which is a combination of previous Acts such

as the Clean Water Act, Water Resources Act, Clean Air Act and other legislation. The objectives of the acts are to protect the environment and public health and to protect farmers from frivolous lawsuits. The EPEA is violated when a significant adverse effect occurs. The Code of Practice covers land use siting and manure management.

The Department of Agriculture recommends an annual maximum of 22–27 tons/ha of solid cattle and hog manure on dryland soils and 56–67 tons/ha on irrigated soils.

British Columbia
The province of British Columbia has the Waste Management Act, the Health Act, the Municipal Act, the Agricultural Protection Act, the Agricultural Waste Control Regulation and the Code of Agricultural Practice for Waste Management. The Code was incorporated in the Waste Management Act in 1992. Its actions are coordinated by an Agricultural Environmental Protection Council (AECP) composed of farmers and governmental authorities, and oversees 150 volunteer farmer inspectors in the Agricultural Environmental Service (AES) trained to investigate and resolve complaints at the farm level. When the local conflict mediation and peer enforcement is not successful the conflict is transferred to the provincial authorities for enforcement action. No prior permitting is required and enforcement is after the fact and more the exception rather than the rule. The Code has the weight of law and a permit is required when in non-compliance. Prevention is achieved through cooperative compliance with standards.

New Brunswick
The Clean Environment Act and the Clean Water Act regulate livestock operations. A Certificate of Compliance is required to receive governmental loans or assistance. The Certificate is issued on the basis of the quality of management, ability to control pollution and adherance to the provisions and recommendations of the guidelines. These guidelines for livestock manure and waste management also contain provisions for setbacks, site planning and manure management.

Nova Scotia
The Nova Scotia Environment Act regulates the waste management practices. The Guidelines for the Management and Use of Animal Manure require regular soil testing to determine the nutrient status, manure testing to establish the nutrient value and spring application and immediate incorporation.

Manitoba
This province relies on regulations rather than voluntary compliance. The Environment Protection Act contains the Livestock Production Operation Regulations and the Livestock Waste Regulations. Manure management is also regulated by the Farm Practices Protection Act and through municipal bylaws and conditional use permits. An engineer's certificate is required when an earthen manure storage

is made. Farms that create disturbances but have acceptable practices are protected from unwarranted nuisance suits. In areas zoned for agriculture small livestock operations are permitted while large units require final permission from the municipal council without appeal.

Ontario

The province of Ontario has the Environmental Protection Act, the Health Promotion and Protection Act, the Planning Act, the Ontario Water Resources Act, the Farm Practices Act and the Agricultural Code of Practice since 1976 (now Guide to Agriculture Land Use). The Code also uses the concept of animal units to provide equivalence and management practices. The compliance with the Code is voluntary. Farmers can obtain a Certificate of Compliance, issued after an inspection by an authorized official. The certificate is also used as assurance for loan organizations to reflect acceptable operations.

Technical provisions are required such as liquid manure storage tanks with a capacity of 200 days and no spreading in the period from January to March. Land spreading should be limited to 75% of crop needs and 20 feet away from water courses. In drained fields the tiles have to be checked for colour change during spreading and should be plugged when that occurs to prevent surface water contamination. Milk house wash water (13.5 liters per cow per day) can be added to the liquid manure, discharged in a septic tank with drainfield or fed to calves. Fencing of cattle out of bottomlands and ditchbanks near streams is recommended to prevent silt contamination and destruction of fish spawning beds (OMAF, 1993)

Prince Edward Island

This province is entirely dependent on ground water for drinking water. Private wells with an average depth of 30 meters provide 70% of the supply. Farm areas comprise 48% of the total land.

The Environmental Protection Act prohibits pollution with no exemptions for farmers. It also addresses waste treatment of hog manure which requires written approval from the Minister. A court injunction can require a person to remedy a contamination.

Quebec

The province of Quebec has the most stringent regulations in Canada for the control of water and air pollution from livestock operations as formulated in the Environmental Quality Act, Regulation Respecting the Prevention of Water Pollution in Livestock Operations (from 1981 as amended in 1984, 1987 and 1990) and Directives for Protection against Air Pollution from Livestock Operations.

Rapid expansion of pig production, without adequate land base, resulted in many incidents of water pollution in rivers and streams. Several municipalities have now forbidden expansion or location of new facilities. Many municipalities require a Certificate of Authorisation to demonstrate regulatory compliance. Air pollution control directives are aimed at controlling odour nuisances and require separation distances of 75 meters for manure storage tanks and 300 meters for manure spreading.

The Quebec Ministry of Agriculture, Fisheries and Food initiated the 'Club Consultation Programme' to encourage sustainable resource management and reduction of non-point sources. Each club or group of about 20 producers sets specific goals for resource parameters and initiates farm-scale conservation projects. An adviser for the project is financed for 90% by the province. This programme is expected to replace the regulatory programme.

Saskatchewan

This province relies primarily on regulation rather than voluntary compliance. The Pollution by Livestock Control Act of 1971 was revised in 1984 and requires a permit for siting and altering livestock facilities. Permits are issued when it is determined that the operation will not cause pollution of surface or ground water and whether adequate provisions are made for manure disposal. The Environmental Management and Protection Act (EMPA) states that the owner of a pollutant is liable for damage that is a result of own neglect or fault.

CHAPTER 11: INTERNATIONAL COMPARISON

In this chapter, an effort will be made to compare the agricultural mineral policies of the different countries, that have been presented in the previous chapters, on a number of different dimensions. The policies that have been developed by the different countries, often differ quite markedly in the ways and means in which they try to achieve their goals, even though their ultimate objective is always the same: the reduction of minerals from agricultural sources in ground- and surface water and a lessening of the nutrient load in the marine waters (in the EU mainly the North Sea). The maintenance of drinking water quality and the protection of surface water against eutrophication are the ultimate goals.

Before a meaningful comparison of policies can be made, however, it is necessary to first compare the environmental pressures that the different countries face relative to the demands of the Nitrate Directive and pressures to realize equilibrium fertilization. A comparison of some of the differences in country conditions is given in Table 1.3.

COMPARISON OF ENVIRONMENTAL PRESSURES

The use of organic fertilizer

When the actual production of nitrogen in fertilizer (or its best estimate) is compared with the norm of 170 kg N/ha utilized in the Nitrate Directive, it becomes clear that in most of the EU countries these norms are not surpassed, while in some other parts the surplus can fairly easily be solved by manure transport to other regions (Fig. 11.1; de Cooman *et al.*, 1995). In some areas, however, the problems are more structural and nitrogen surpluses exceed 200 kg N/ha (the Netherlands), while in Belgium (Flanders), the regional surplus exceeds 100 kg N/ha. These countries have a serious mineral surplus, that has been stabilized in recent years but has not been reduced very much. Parts of Germany (Münster, Weser-Ems, Düsseldorf) and France (Bretagne) have high manure production levels combined with high fertilizer use, resulting in a mineral surplus.

In these countries and in Denmark, legal limits have been placed on animal production, which might lead to a reduction of the pig production sector. Manure transports are not taken into consideration in these figures, but manure transports alone are inadequate to sufficiently reduce regional surpluses in the Netherlands and Flanders. Action programs, that have been introduced, in combination with Codes of Good Agricultural Practice, will have to work towards a further reduction. Much will depend on the political will of the governments involved, whether really serious reductions will be realized, or if they will remain marginal.

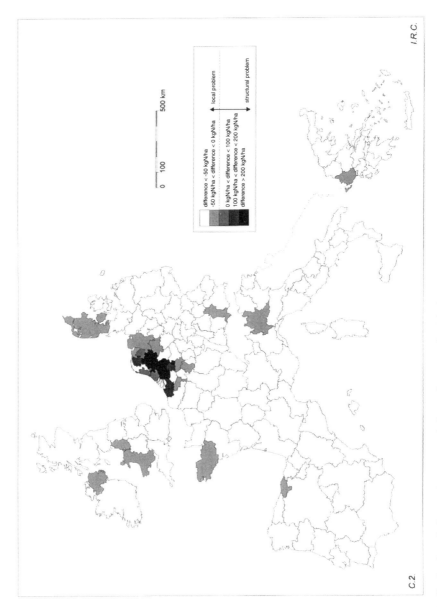

Figure 11.1 Production of nitrogen in manure relative to the norm of 170 kg/ha. Source: de Cooman *et al.*, 1995

The use of mineral fertilizer
When the actual use of mineral fertilizer (or best estimates thereof) is compared with the norms that are part of the Code of Good Agricultural Practice for each country, it becomes apparent that in some parts of Germany, Denmark, France and the UK the fertilization exceeds the norm (Fig. 11.2; de Cooman *et al.*, 1995), although not to any great extent. In fact, the authors point out that in those areas where agriculture is intensive, but with little or no livestock farming, there is very little difference between the norms and recommendations that are part of the Code and actual agricultural practice. This means that these agricultural practices try to achieve an economic optimum and are fairly successful in that respect.

Equilibrium fertilization
One objective of the Nitrate Directive is the use of equilibrium fertilization, in which the level of fertilization, both organic and mineral, does not exceed the needs of the crops. However, it is difficult to 'translate' this concept into the Code of Good Agricultural Practices, and countries find very different solutions to this problem.

When the actual levels of manure production and mineral fertilizer use are compared to the limits, as they are laid down in the Codes, it becomes necessary to make a lot of estimates and figures become very approximate. Nevertheless, Cooman *et al.* (1995) have endeavoured to make a close approximation of this difference and the result is presented in Fig. 11.3. This figure shows that the regions with intensive livestock operations correspond closely to the designated Vulnerable Zones: the whole of the Netherlands, Germany and Denmark are designated as such, along with Flanders in Belgium, Bretagne in France, and smaller areas in the UK. In these areas, it will depend on the results of the action programmes whether sufficient reductions can be achieved within a reasonable time frame. The ultimate effects of the programmes will depend to a large extent on the political will of the countries involved to implement and enforce action programmes that are sufficiently stringent to really reduce mineral surpluses.

The results of a study which calculated regional mineral balances for the EU, based on both mineral and organic fertilization, is shown in Table 1.2 and Fig. 3.1 (Scheef and Kleinhanss, 1996). These authors utilize more and smaller surplus categories and are thus able to make finer distinctions between regions, but nevertheless the correspondence between Figs 11.3 and 3.1 is apparent.

Direct losses and run-off
Measures A1–A5 of the Code are intended to reduce direct losses and run-off of minerals due to inappropriate methods of manure application. An estimate was made about the potential effect of these measures on the nitrate concentration in surface water (de Cooman *et al.*, 1995). The most significant results of these measures were to be expected in areas with intensive livestock farms. However, in those areas these measures have already been implemented for a long time. This means that the additional effects that could still result from these measures in

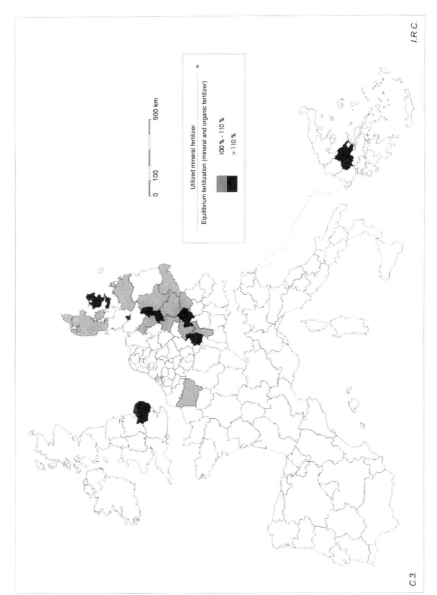

Figure 11.2 Use of mineral N fertilizer relative to the norm of an equilibrium fertilization. Source: de Cooman *et al.*, 1995

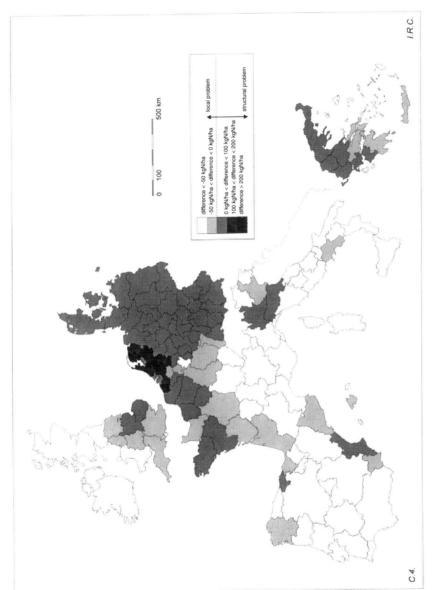

Figure 11.3 Application of N fertilizer relative to the norm of an equilibrium fertilization. Source: de Cooman et al., 1995

those areas are probably not that significant. Some impact is still to be expected in countries that are now implementing these measures. Since these areas have less intensive livestock farming, the effects will not be that great.

The USA and Canada

Conditions in the USA and Canada differ a great deal from those in the EU. Land is not a limiting factor and in the dry western states, livestock operations tend to be very extensive. In these areas, problems can be overgrazing. Trampling and defaecating in riparian strips bordering pristine streams is seen as a major problem for habitat destruction and drinking water impairment, requiring management practices.

In some Midwestern states, such as Ohio, Minnesota, Ontario and Alberta, there are large feedlot operations with more than 1000 animal units requiring the collection and treatment of liquid waste streams. These operations are very intensive. The nutrient production in the different states generally does not exceed crop needs so that over-applications on the land only take place in areas with very intensive livestock operations. For these operations, manure and run-off treatment requirements are comparable to approaches for industrial wastewaters.

COMPARISON OF POLICIES

The comparison between the policies of different countries will be made on a number of different dimensions, in order to be able to highlight the differences or emphasize their similarities.

Policies that limit the production of manure

In the middle of the 1980s, both Denmark and the Netherlands opted for policies to put a ceiling on the total amount of manure produced in each country, but they have chosen different ways of doing so.

In the Netherlands, the Manure and Fertilizer Act was adopted in 1986. The Act prohibits the establishment of new holdings or the expansion of existing farms above the point where the production of phosphate in manure exceeds 125 kg/ha per year. Farms then have 'manure production rights', which limit their expansion. These rights will be siphoned off by the government when the holding is sold outside the family (25%). For a lot of farms that were already in existence at that time, the manure production rights exceeded the limit of 125 kg/ha. The expansion of manure production did stabilize at that point, however.

Belgium adopted a similar rule to create a standstill of phosphate production in manure in 1992. This means that total phosphate production can not expand beyond the 1992 limits. Municipalities in Flanders are classified into four classes, some of which allow expansion, where in others farmers can only expand if others reduce their production. Part of the manure production rights can be siphoned off by the government under certain conditions.

In Denmark, the Action Plan for the Aquatic Environment was adopted in

1987, which includes the Harmonization Rule. This rule sets upper limits for the application of manure: these limits are 2.3 LU/ha on dairy farms and 1.7 LU/ha on pig farms. Farms with higher densities may comply with these standards by agreements with other farmers to receive their surplus manure. In this manner, Denmark has also capped manure production, but has done so by requiring a fixed ratio between manure production and the amount of land it can be applied to.

The Netherlands and Belgium have thus limited the mineral production at a certain level, while Denmark requires a fixed ratio between livestock density and land area. Apparently, this rule has motivated some Danish farmers to move their intensive livestock operations to Poland, reducing mineral supplies in Denmark (but at the same time compounding environmental pressures in Poland). When comparing Denmark on the one hand and the Netherlands and Belgium on the other, it seems that, 10 years later, Danish policy has been more effective in keeping mineral production within bounds than the Dutch policy has. The Danish livestock sector is much better able to utilize its manure in keeping with environmental demands.

The Dutch policy does enable the government to 'siphon off' manure production rights and there have been proposals to do so at a higher rate in the future, but these proposals have met with a lot of resistance from farmers and have been rejected. Thus, the government does have a policy instrument that allows it to reduce and restructure the intensive livestock sector, if it decides to do so, but this issue is still too sensitive politically. The Belgian policy has not been in existence very long, but the situation is similar to the Netherlands.

In summary, one may conclude that the Danish Harmonization Rule has been more effective than the Manure Production rights, but adopting it in the Netherlands and Belgium is no longer an option: the ratio between manure production and available land in these countries is already too lopsided. Either a significant reduction of the livestock population or some other fairly drastic measures will be needed to realize some 'harmony' in these countries.

Policies limiting the application of minerals in manure and mineral fertilizer
The Nitrate Directive has put a fixed limit on the application of animal manure: 170 kg N/ha per year, while allowing a limit of 210 kg N/ha per year until 1999. In addition, the Directive aims to stimulate 'equilibrium fertilization', where the total of minerals applied, from both mineral and organic sources, does not exceed the needs of the crops, while allowing for a certain inevitable loss. However, how equilibrium fertilization is to be implemented is not uniformly specified. Thus, different countries have found different policy options for implementing it.

The Netherlands, in adopting the Minerals Accounting System (MINAS) as compulsory on intensive livestock farms in 1998, has opted for a policy that more or less bypasses the fixed limit of 170 kg N/ha. MINAS is an individual instrument to help farmers achieve equilibrium fertilization, while taking into account both organic and mineral fertilizer and utilizing norms for both P and N. It also takes into account the uptake of different crops. Negotiations with the EU are still ongoing about the question whether MINAS is an acceptable alternative to the fixed norm of 170 kg N/ha.

Denmark requires fertilizing plans for both mineral and organic fertilizer, while specifying a minimum utilization rate for animal manure. Belgium (Flanders) and Germany also adopted maximum acceptable loss norms, while accounting for the total amount of N produced in animal manure. Other countries put a limit on the use of mineral fertilizer, without accounting for animal manure (Belgium/Wallone, France, UK).

It seems that the Dutch policy is well suited to the requirements of an equilibrium fertilization, while allowing farmers maximum control over their own management options. Important to the effectiveness of this policy, however, will be the level of the acceptable loss norms that are decided on and the level of the levy on surpluses. If norms are too high and levies are too low, they will lose their effectiveness (Hellegers, 1996). Recent decisions to lower levies in the next two years and make the loss norms less stringent than initially proposed, will reduce its effectiveness.

The Code of Good Agricultural Practices
The compulsory part of the Code (A1–A6) is similar for all countries. Countries with an intensive livestock sector have made most requirements into law, while in other countries they are advisory in nature. Nevertheless, several of these measures are difficult to enforce and control, but this is true for all countries involved.

Command and control versus market-driven policies
Another distinction that can be made between policies is the command and control type policy versus market-based ones. Marked-based policies have the advantage of offering financial incentives to farmers to make choices that comply with government policy, without being dependent on enforcement from the side of the government, as is the case with command and control policies.

In practically all countries, agricultural mineral policies are of the command and control type. This also creates a lot of problems, because the agricultural sector consists of many, relatively small, independent operators and control and enforcement of regulations becomes very time consuming, while some rules are very hard to control at all. Command and control type regulations also generate a lot of resistance, thus rendering policies less effective, while farmers' protests, judicial procedures etc. can demand a lot of time and energy from authorities.

The Netherlands has recently created the possibility for rewarding farmers who stay below fertilization limits with premiums, while maintaining levies on surpluses. This would provide a better stimulus to reduce mineral loads beyond official limits than just avoiding the levy. A system, that is more market-based would be to allow 'tradeable emission rights' for manure and fertilizer, where farmers who stay well below the official norm, are able to sell that part of their emission rights to others.

Another market-based policy would be to levy a tax on mineral fertilizer. This proposal was initially accepted in Denmark in the late 1980s, but never became law, because of political opposition. In Canada, the province of Ontario has a certification procedure, dependent on compliance with the Code of Practice, that enables farmers to obtain better insurance rates and more favourable loans. Rewarding compliance behaviour, if carefully structured, may offer a more efficient alternative or addition to the command-and-control type of policy.

North American policies

In the USA, regulations are primarily focused on large feedlot operations with more than 1000 animal units, where collection and treatment of liquid waste streams are required. The Federal program is mostly delegated to the individual States, who in turn delegate portions to their counties. There are no policies to limit the size of the animal operation, thus allowing economies of scale. Manure and run-off treatment requirements are comparable to approaches for industrial waste waters. Thus, for large operations the USA has clearly chosen an industrial approach to manure processing and treatment. This industrial approach to livestock waste processing may present a viable alternative for other countries with very intensive livestock farms and high mineral surpluses. Large holdings with high livestock densities and very little land for the disposal of animal wastes could be required to treat their waste streams to meet standards comparable to those used for industrial plants.

The environmental management provided in Canada is primarily directed towards surface waters and fish protection with substantial fines for violations. Most regulations for manure management are voluntary.

Extension services, research and voluntary inspectors

Another alternative to command and control type policies are extension education programs, usually in combination with research efforts, to find scientific ways for reducing mineral surpluses and promoting their adoption. Both England and France have emphasized this alternative. In the Netherlands, the Extension Service is actively involved in these efforts.

The province of British Columbia has a volunteer farmers inspectors programme based on peer enforcement and conflict mediation, with backup from legal authorities.

CONCLUSIONS FROM INTERNATIONAL COMPARISON

The comparison between environmental pressures and mineral agricultural policies of the eight countries involved shows that environmental pressures are the greatest in the Netherlands and Flanders, where considerable surpluses exist, because both manure production and mineral fertilizer use are high. Similar problems exist in parts of Germany and France and the UK, where problems are

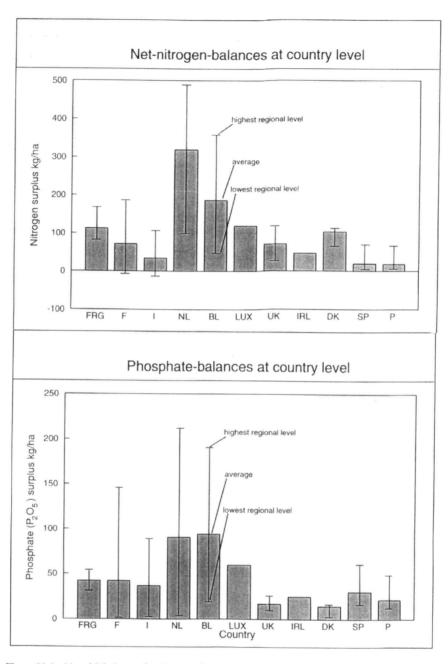

Figure 11.4 N and P balances for 11 countries. Source: Schleef and Kleinhanss, 1994

more regional and in Denmark, where the level of the surplus is more manageable.

Denmark's Harmonization Rule has been more effective in limiting manure production by maintaining a ratio between land area and manure production than the manure production rights, utilized by the Netherlands and Belgium. In the latter countries, manure production has stabilized, but at a level that still exceeds demand for manure.

The Mineral Accounting System (MINAS) that is being introduced in the Netherlands, offers farmers more individual control over manure production and disposal than systems that use a fixed application norm. However, this system still has to be accepted by the EU.

The USA and Canada have adopted a more industrial approach to large feedlot operations, that have to meet the same treatment and discharge requirements that industrial plants are held to. This approach may present a viable alternative for European countries with a very intensive livestock sector. Canada uses voluntary compliance programs and volunteer inspectors to enforce norms.

Market-based policies may be more effective and will require less enforcement than the command and control type policies that are used by most countries. Their further development could gradually replace command and control type measures. Figure 11.4 also illustrates the difference in mineral balances between EU countries.

CHAPTER 12: CONCLUSIONS AND RECOMMENDATIONS

Deterioration of the environment as a result of the emission of minerals from agriculture has created increasingly serious problems in several European countries. Nitrate pollution has become a serious political issue in the European Union. The objective of the EU Nitrate Directive of 1991 is the protection of ground- and surface waters against nitrate pollution from agricultural sources. An important element of the Directive is that the application of animal manure should not exceed 170 kg N/ha in designated zones, that are vulnerable to the leaching of nitrates. In addition, member states have to designate vulnerable zones and set up an action programme specifying policies and measures they will take.

In this study, an overview was presented of the agricultural policies on manure and minerals in six different EU countries, especially those relating to the Nitrate Directive. These policies and their legal incorporation were related to agricultural and environmental conditions in each country. In addition, an inventory was made of agricultural mineral policies in the United States and Canada. Conditions for livestock farming in North America differ considerably from those in Europe, but their solutions may shed a new light on European policies. Summaries will be presented for each country, after which conclusions and recommendations will be made. The policies to stop run-off and leaching of manure and silage and those to ensure adequate storage capacity for manure, as formulated in the Code of Good Agricultural Practice, have been adopted, with some variations, by all six EU countries and will not be mentioned separately, but those policies that are unique for each country will be mentioned.

THE NETHERLANDS

The Netherlands is a small, densely populated country with a highly developed agricultural sector. Livestock farming is very intensive and livestock densities are high, especially on the sandy soils in the eastern, middle and southern parts of the country. The production of organic fertilizer is high: the average for the whole country is 339 kg N/ha and varies from a low of 63 kg N/ha to a high of 609 kg N/ha. On 99% of all farms, where manure is produced, manure production exceeds the EU limit of 170 kg N/ha. In addition, mineral fertilizer use is also very high compared with other European countries, averaging 218 kg N/ha for the whole country. Consequently, the nitrogen balance for the Netherlands as a whole is 321 kg N/ha, the highest level in the EU. As a result of this considerable mineral surplus, phosphate saturation of sandy soils has become a serious problem and virtually all surface water is at risk of eutrophication.

In the mid-1980s, when the need for more environmental regulation became apparent, the Dutch government decided that drastic changes would have to be made, but that it would be impossible to do so overnight. Consequently a gradual change process was envisioned in three phases, each with their own objective. In the first phase (1987–90), the objective was to stabilize manure production at a level where no national surplus would occur. Manure production rights and application standards for livestock manure were introduced. Manure banks were organized, to redistribute excess animal waste. In the second phase (1990–94) application standards for manure were tightened. In the third phase (1995–2000) the policy goal of an equilibrium between the input of mineral and organic fertilizer and the output of harvested crops will have to be achieved. The manure application standards are being replaced by a system of 'acceptable loss norms'. In order to realize these loss standards at farm level, a new accounting system (MINAS) will be introduced, that will allow individual farmers more control over input and output. MINAS will become obligatory for some farms in 1998, but levies will be low for the first 2 years, diminishing its effectiveness. The acceptability of MINAS for the EU Nitrate Directive is still being negotiated. The whole country has been declared a vulnerable zone. Research shows that application rates for both P and N in manure show moderate reductions between 1985 and 1992 (29% for N, 16% for P). For mineral fertilizer these figures are 23% for total-N and 11% for total-P. Recent data show that in 1994 the N-surplus increased by 2%, in 1995 a 5% increase is estimated. The objective of a 50% reduction in 1995 has thus not been realized and no major reductions are to be expected in the near future.

BELGIUM

Belgium is a federal state composed of three regions. Wallonia has no significant manure surpluses: farms are mostly cropping and dairy farms. In Flanders, however, livestock farming is very intensive, especially pig and poultry farming, and livestock densities are high. In the Brussels area, agriculture is of minimal importance. The average production of organic fertilizer is 300 kg N/ha for Flanders, while in Wallonia it is 145 kg N/ha. For the province of Antwerp, the average is 425 kg N/ha. The mineral balance for the whole country is 178 kg N/ha, but is much higher for some Flemish provinces: 358 kg N/ha for Antwerp, 308 kg N/ha for West-Vlaanderen, while for Wallonia it is much lower. Thus, the situation in Flanders is comparable to that in the Netherlands, even though manure surpluses are somewhat lower. In this part of the country, almost all surface waters know some degree of eutrophication. About 40% of the drinking water from private wells exceeds the drinking water standard.

The Environment and Nature Plan (MINA Plan 2000) for Flanders outlined policies for 1990–1995, consisting of two parts: an inventory of the environmental situation and a number of action plans. The Manure and Fertilizer Decree (1991) set fairly liberal fertilizing limits and set up a Manure Bank for the distribution of excess manure. In 1992 a stand-still in manure production was declared. The total quantity of P and N that could be produced in livestock manure was frozen at the 1992 level. Possibilities of expansion can only be realized in municipalities with room for expansion. The new MINA Plan 1997–2002 aims for equilibrium fertilization. The New Manure and Fertilizer Decree (1996) sets stricter limits on all fertilizer use and sets a levy. It requires manure accounting and sets specific limits for protection areas. It also introduced the concept of the family livestock farm, which is positively discriminated in trading and transporting their manure surpluses. All of Flanders has been declared a Vulnerable Zone for the EU Nitrate Directive. Research shows that average manure application rates in Belgium have increased somewhat between 1985 and 1992, for both P and N, while other estimates show moderate expected reductions for both minerals.

DENMARK

Denmark is a small country, about the same size as the Netherlands (4 million ha). Its livestock population consists of 4 million LUs, about half of the Dutch livestock population, and consists of 56% of pigs, 38% cattle and 5% poultry. Most of its agricultural area (93%) is arable land. The average production of organic fertilizer is fairly low at 102 kg N/ha for the whole country and not much higher in different regions. At farm level, 26% of all holdings have supply levels that exceed the limit of 170 kg N/ha and their average supply of N is 258 kg/ha. This means that livestock production is very intensive on a small percentage of farms. Mineral fertilizer use is in the middle range compared with other EU countries (140 kg N/ha). The nitrogen balance for the whole country is 104 kg N/ha and does not vary much across regions. Drinking water is still of good quality in Denmark. The level of nitrogen in marine waters has stabilized in the late 1980s, but eutrophication is still a serious concern.

   Marine water quality is very important in Denmark, both economically and for recreation. For that reason, concerns about water quality deterioration struck a responsive chord, both politically and in the public opinion. This has resulted in far-reaching policy goals and the most comprehensive abatement programme in Europe. Three major Action Programmes were implemented between 1985 and 1992. The NPO Action Plan was accepted in 1985 and replaced in 1987 by the more comprehensive Action Plan on the Aquatic Environment. A very important element in this plan is the 'harmonization rule', which matches herd size with the land available. This means that a sufficient area must be available for spreading of farmyard manure. If the area belonging to the farms is insufficient, long-term lease contracts with other farms for the disposal of manure are accepted. This rule has prevented very high livestock densities on farms and the development of intensive livestock farms that are independent of land area, as has happened in the Netherlands and Flanders. In 1991, the Programme for a Sustainable Agriculture was introduced: its objective is to realize a better utilization of farmyard manure, by setting minimum utilization percentages and ties the total amount of nitrogen applied to the estimated need for N on the farm. This requires fertilizer and crop rotation plans from farmers. In Denmark, winter crops are frequently used and this practice is increasing. In 1988, the Nationwide Monitoring Programme (NWMP) was set up, to monitor the effects of programs and policies. A proposed tax on nitrogen in mineral fertilizer was first accepted, but has not been implemented due to political opposition. The whole country has been declared a vulnerable zone. The Ministry of the Environment has estimated that nitrogen emissions from agriculture were reduced by 20% between 1987 and 1990. PARCOM data, on the other hand, show small reductions in N and P from both manure and fertilizer; OSPAR studies show a 15% reduction for N and 5% for P between 1985 and 1995.

FRANCE

France is the biggest country in Western Europe. French agriculture is very diverse, with considerable differences among the different regions. It has the largest livestock population in the EU, consisting mostly of cattle (70%), pigs (12%), poultry (10%) and sheep and goats (6%). Bretagne and Pays de la Loire have the largest livestock populations (4.8 and 3.1 million LU respectively). Only in Bretagne is livestock farming very intensive, with densities of more than 2 LU/ha. Manure production in Bretagne is 150 kg N/ha, but much lower in all other regions. The average manure production for France is 52 kg N/ha. Average mineral fertilizer use is 108 kg N/ha in Bretagne, 91 kg N/ha for the whole country. The average nitrogen balance for France is 63 kg N/ha. Bretagne has the highest nitrogen balance at 130 kg N/ha, related to intensive livestock farming. In Bretagne, some aquifers are very polluted by nitrates. High nitrate levels in the coastal waters of Bretagne-Cotes d'Armor have caused serious eutrophication outbreaks, that interfere with tourism. Even though the intensity of livestock farming in this area is not as high as in the Netherlands and Flanders, it has caused a lot of pollution problems because farms and storage facilities are often old and in disrepair.

The publication in 1980 of the Henin report about the relationship between agricultural activity and water quality, started the public debate in France about nutrient pollution. As a result of this report, a new administrative body, CORPEN, was set up to implement measures and coordinate the campaign against nitrate pollution. Its executive body is the Mission Eau-Nitrates or MEN. In general, the role of CORPEN and MEN is to promote research studies related to nitrate pollution control. They operate on a consensus strategy, which aims at reconciling high performance in agriculture with water quality. Action should be achieved through incentives rather than through constraints. The implementation of the Nitrate Directive is legally incorporated in the 'Decree of 27 August 1993'. The Decree specifies how to define vulnerable zones, based on the average level of nitrate concentration during the past year. Waters with nitrate levels over 50 mg/l are considered vulnerable. It also defines the Code of Good Agricultural Practice. An Interministerial Agreement between the Ministry of the Environment and the Ministry of Agriculture was made with the objective of introducing a polluter-pays principle into agriculture. In this agreement, priority is given to preventive measures such as research and extension programs; economic incentives like subsidies and levies are also part of it. PARCOM data show no changes in manure application; mineral fertilizer application rates increased for both P (19%) and N (13.5%) between 1985 and 1992. Provisional OSPAR estimates give reduction rates for P and N of 17% and 10%, respectively, in the same period.

GERMANY

Reunification made Germany one of the larger countries in the EU. It is a federal republic, consisting of 13 'Länder' (states), that are subdivided into counties and three city-states. The federal republic defines the general legislative framework, which then has to be implemented in legislation by each of the Länder. The agricultural sector in Germany is large and varied across regions. Only Niedersachsen and Nordrhein-Westfalen have some counties with livestock densities over 2 LU/ha. In the new Länder, livestock farming used to be very intensive before unification. Presently the average livestock density is between 0.9 and 1.2 LU/ha. The average production of livestock manure for the whole country is 82 kg N/ha. The highest levels are found in Münster (151 kg N/ha) and Wester-Ems (146 kg N/ha). These areas have the highest livestock densities and the highest concentration of pig farms. Research at farm level shows, that manure production exceeds the 170 kg N/ha limit on only 12% of the holdings. The use of mineral fertilizer is 128 kg N/ha, with a high of 175 kg N/ha for Braunschweig. Mineral fertilizer use increased between 1970 and 1980, then stabilized and since 1990 it has shown a distinct drop. The average nitrogen balance for Germany is 65 kg N/ha. The counties with the highest surpluses are Münster (167 kg N/ha), Wester-Ems (155 kg N/ha) and Düsseldorf (146 kg N/ha), but none exceeds 170 kg N/ha. Groundwater quality is a major concern in Germany. Sudden breakthrough of nitrate, which has been observed in several water extraction works, is problematic. The drinking water quality standard of 50 mg NO_3/l is exceeded in about 20% of the groundwater stations in Niedersachsen and Nordrhein-Westfalen. Marine water is threatened by eutrophication, especially in the Deutsche Bucht.

In Germany, environmental organizations have been influential and politically active (the Green Party) for a long time. In the late 1970s, surface water quality was the main concern: nitrate pollution of groundwater is currently a more prominent political issue. For the Nitrate Directive, the whole country has been declared as a vulnerable zone. In 1996 the new Fertilizer Decree was approved, in which the Code of Good Agricultural Practice has been codified. It supplants fertilizer regulation that several states had enacted on the basis of the waste laws. It decrees that enterprises with more than 10 ha of agriculture have to provide a mineral balance accounting. PARCOM data show that manure application of N increased by about 18% between 1985 and 1992, while P application decreased by almost 30%. Mineral fertilizer application decreased 25% for total-N, 58% for total-P. The total inputs of nutrients from agriculture show an expected reduction of 21% for P, 17% for N between 1985 and 1995 (OSPAR).

UNITED KINGDOM

The UK consists of England, Wales, Scotland and Northern Ireland. Of the total area 78% is agricultural land, mostly permanent grassland. There are marked differences between regions. The livestock population consists for an important part of sheep and cattle. Pigs are mostly held in Yorkshire, Humberside and East Anglia, sheep and cattle are important in Scotland, Wales and the south west. Livestock farming is not very intensive; livestock densities do not exceed 2 LU/ha anywhere in the UK. The average manure production for the whole country is 66 kg N/ha. The highest level of manure production is found in north west England (132 kg N/ha) and Northern Ireland (124 kg N/ha). The average level per region does not exceed 170 kg N/ha anywhere. Average mineral fertilizer use was 130 kg N/ha for the whole UK, with a high of 167 kg N/ha in East Anglia. The nitrogen balance for the whole country is 80 kg N/ha. The nitrogen balances are highest in Cheshire (117 kg N/ha), a county with an intensive dairy sector, and Humberside (109 kg N/ha), with a high density of pigs in proximity to arable land. The positive nitrogen balances are the result of a combination of stock numbers and fertilizer use. The concentration of nitrates in surface waters increased steadily in the period 1970–1985. Since 1985, nitrate levels seem to have stabilized. The increase in nitrate level is not only due to nutrient loss in agriculture, but is also related to domestic waste from sewage treatment plants and untreated sewage (about 17% in 1992). A major investment program is now in place to remedy this problem. In the UK, about 70% of the drinking water supply is abstracted from surface water. Groundwater sources are found in the drier eastern and southern regions, where a significant part of the aquifers is vulnerable to nitrate leaching for geological reasons. In this part of the country, arable farming predominates. The UK differs from other EU countries in that high nitrate levels are more related to high mineral fertilizer use, rather than livestock manure, especially in the southeast. The quality of coastal and marine waters in the UK has not been a problem: the prevailing northern currents in the North Sea favour the UK by diluting and dispersing polluting substances away from the coast.

In the UK, nitrate pollution is perceived as mostly a drinking water problem. Eutrophication of coastal waters and rivers is not considered a major problem. The UK has accepted the Nitrate Directive standards for drinking water only after some controversy and has accepted the PARCOM recommendations only partially. Under the Environment Protection Act of 1991, 32 Nitrate Sensitive Areas (NSA) have been designated (within the vulnerable zones) where farmers voluntarily make substantial changes in their farming practices to reduce nitrate leaching, in return for compensating payments under the EC Agri-environment regulation. As a result of the Nitrate Directive, 69 Vulnerable Zones have been designated, covering about 5% of the agricultural land. They will become operable in 1999 and will include the present NSAs. The Code of Good Agricultural Practice is advisory in nature, even though it contains some legal requirements. Under the Environment Act of 1995, two unitary pollution control authorities have been set up, one for England and Wales and one for Scotland. They have the power to control pollution to any of all three media: water, air and soil. The UK has emphasized research and extension education in diminishing nutrient application. PARCOM data show that the application rates of manure have not changed much between 1985 and 1992. The application rates of mineral fertilizer show a 14% reduction for total-N and a 17% reduction for total-P between 1985 and 1993.

THE UNITED STATES

The US regulations are primarily focused on large feedlot operations: those with more than 1000 animal units require the collection and treatment of the liquid waste streams. The Federal programme is mostly delegated to the individual States, who in turn delegate portions to their counties. There are no policies to limit the size of the animal operation, thus allowing large economies of scale. The nutrient production in the different states generally does not exceed crop needs so that over-application on the land only takes place in intensively used areas. There is no effort to favour the land application of manure; in fact, discharge into surface waters after treatment is very common. Treatment requirements are comparable to approaches for industrial waste waters. Thus for large operations the USA has clearly chosen an industrial approach to manure processing and treatment.

In the dry western states overgrazing, trampling and defaecation in riparian strips bordering pristine streams is seen as a major problem for habitat destruction and threats to drinking water supplies, requiring management practices. Such measures related to river edge strips were not encountered in Europe.

CANADA

The environmental protection provided in Canada is primarily directed towards surface waters and fish protection with substantial fines for violations. Most regulations for manure management are voluntary. The province of Ontario has a certification programme for complying farmers. The province of British Columbia has a volunteer farmer inspectors programme based on peer enforcement and conflict mediation, with backup from the legal authorities.

CONCLUSIONS AND RECOMMENDATIONS

One conclusion that becomes very apparent in this study is that there are still very considerable mineral surpluses in some countries, that are not regional and therefore cannot easily be solved with manure transports to other regions. In all countries that have designated their whole territory or a large part of it as vulnerable zone, serious problems still exist. In both the Netherlands and in parts of Flanders, existing problems are clearly structural rather than local. To a lesser extent, Germany, Denmark and relatively smaller parts of France (Bretagne) and the UK still exceed the norms for an equilibrium fertilization. Denmark is in a position where it can probably solve its problems within the existing legislative framework, but the Netherlands, Belgium (Flanders), some German Länder and Bretagne, because of the intensity of their livestock sector, would require more structural solutions. In these countries or regions, nutrient surpluses have been stabilized and moderate reductions have been achieved with the existing measures. The action programmes do comprise new measures to realize further reductions, but in most cases, the political will to enforce stringent regulations and impose levies is not there. Thus, it remains to be seen whether the action programmes will be able to sufficiently reduce the present surpluses or whether more structural solutions need to be found.

In the remaining countries, that is France (except Bretagne), the UK and Belgium (Wallon), mineral problems are more local. The Code of Good Agricultural Practices has been accepted on a voluntary basis and some legal regulations exist. Action programmes apply to limited areas that have been declared vulnerable.

The evaluation of the possible impact of the Code of Good Agricultural Practice shows that the measures A1–A5 that prevent direct nitrogen losses to surface water, will have an effect in countries that are still implementing these measures. In countries where the agricultural mineral problems are more serious, they have already been implemented and have contributed to a stabilization of the nitrate concentration in surface water. In these countries, more significant effects are to be expected from the Action Programs, that are based on equilibrium fertilization, even though the effectiveness of these programs is still uncertain and depends to a large extent on the willingness of authorities to enforce these measures.

A comparison of policies that the different countries have adopted shows that Denmark, by maintaining a fixed ratio between livestock density and land area, has been able to prevent the livestock sector from further intensification: by adopting the 'harmony rule', livestock farming in Denmark is still dependent on land area. Farms with very high livestock densities and little or no soil, as exist in the Netherlands and Flanders, have either been forced to move abroad or have had to make contracts with other farms, where they can dispose of their surplus manure. The Netherlands and Belgium (Flanders) have been able to stabilize their livestock density and manure production by adopting a policy of 'manure production rights', but at a level that still creates a structural mineral surplus problem, even when manure is transported to other parts of the country. Holdings with very high livestock densities, especially pigs and poultry, still exist, with little or no land area to dispose of manure. These holdings are more industrial than agricultural and are similar to 'feedlots' in the United States and Canada. These North American feedlots are subject to the same treatment requirements for waste water as industrial plants are. Thus, the United States has chosen an industrial approach to manure processing and treatment.

The Netherlands has made a choice for a mineral accounting system, with maximum allowable losses, which is geared towards equilibrium fertilization. It allows farmers more individual control and does not use a fixed limit as is required by the Nitrate Directive. Its effectiveness in reducing mineral surpluses will depend on the fertilization limits that are used and the level of the levies and the degree of enforcement. Other countries use fixed limits or combinations of both systems and their effectiveness also depends on the ability of the authorities to enforce them.

All EU countries rely heavily on command and control type policies. These type of policies are very expensive and time consuming and often meet with a lot of resistance from farmers. They require a lot of control and enforcement and are susceptible to fraud. Thus they are often less effective than they seem to be on paper. Reward- or market-based systems avoid some of these problems.

USA policies are mainly focused on large feedlot operations; treatment requirements for these are comparable to approaches for industrial waste waters. Several provinces in Canada use voluntary regulations and volunteer inspectors, but also fine violators of environmental regulations quite heavily.

Based on the results of this study, several recommendations are formulated. In those countries where agricultural mineral surpluses are structural, a choice will have to be made between a drastic reduction of the intensive livestock sector or the development of other alternatives. One alternative may be to develop policies to differentiate between large operations and small ones. The approach for large operations could be similar to that employed for industrial wastewater treatment using best available control technology with surface water discharge of the effluent and no requirement for land disposal of nutrients. The treatment must not only remove organic materials but also a major portion of the N and P through nitrification/denitrification and luxurious P uptake or chemical precipitation. The resulting sludges are handled similar to those from municipal sewage treatment plants. The size of the operation would not be limited so that economies of scale are realised. This is similar to the approach used in the USA for large feedlots. Smaller farms and livestock operations could follow the example in Belgium or Denmark with a strong link between the size of the operation and the amount of available land for disposal of the manure. This could be the model family farm or biological farm respecting nature conservation and habitat rehabilitation. The above differentiation could start as a command and control operation and gradually move to voluntary measures, as seen in several Canadian provinces.

The above dual system could become more efficient with strong adoption of market-based policies, like tradeable mineral emission rights, heavy taxation of mineral fertilizers, certification and lower insurance rates and waiving of liabilities. Thus, a policy approach is recommended, that differentiates between large, intensive, industrial type operations, that no longer are dependent on land area, that would be subject to industrial waste treatment norms and small, family-type less intensive livestock holdings, with sufficient area (or contracts with other farms) for their own manure disposal.

BIBLIOGRAPHY

Alberts, E., Neibling, W.H. and Moldenhauer, W.C. Transport of sediment nitrogen and phosphorus through cornstalk residue strips. Soil Sci Soc Am J 45, 1177 (1981).

Asselin, R., Madramootoo, C. Effects of agriculture on the quality of the Quebec waters. National workshop proceedings on agricultural impacts on water quality. Canadian Agricultural Research Council, Ottawa (1992).

Baldock, D. and Bennett, G. Agriculture and the Polluter Pays Principle; a study of six EC countries. Institute for Environmental Policy, London, 1991.

Bauer, S.B. and Burton, T.A. Monitoring protocols to evaluate water quality effects of grazing management on western rangeland streams. US Environmental Protection Agency Report EPA 910/R-93-017, Washington, DC (1993).

BMELF. The new fertiliser decree. Federal Ministry of Food, Agriculture and Forestry, referat 312, Bonn (1996).

BMU. Agricultural measures to reduce nutrient input into waters. Bundesministerium für Umwelt, Bonn (1993).

Brouwer, F.M., Godeschalk, F.E., Hellegers, P.J.G.J. and Kelholt, H.J. Mineral balances at farm level in the European Union. Agricultural Economics Research Institute (LEI-DLO), Onderzoeksverslag 137, The Hague (1995).

Brouwer, F.M., Godeschalk, F.E. and Hellegers, P.J.G.J. Mineral balances at farm level in the European Union. Conference Proceedings on the Nitrate Directive and the Agricultural Sector of the European Union. The Hague, June 20–22, 1996.

Canter, L.W. Nitrates in groundwater. CRC Lewis Publishers, Boca Raton (1997).

CEC, Commission of the European Communities, 'The Implementation of Council Directive 91/676/EEC concerning the protection of waters against pollution caused by nitrates from agricultural sources,' Report of the Commission to the Council and European Parliament, COM(97)473def., Brussels, 1997.

CFI. The role of fertilisers in sustainable agriculture and food production. Canadian Fertiliser Institute, Ottawa (1990).

de Cooman, P., Donnez, M. Scokart, P. Les Codes de Bonne Pratique en relation avec la Directive Nitrate, Final Report, Commission of the European Communities, Contract B4–3040/93/00/209, Brussels (1995).

Fong, N. Mineralen in de landbouw, 1994. Kwartaalberichten Milieu (CBS), 4 (1996).

Frederiksen, B.S. Legislation in response to the EU Nitrate Directive – aspects for selected countries. Conference proceedings on the Nitrate Directive and the Agricultural Sector of the European Union. The Hague, June 20–22, 1996.

Frederiksen, B.S., Schou, J.S. Mineral emission from agriculture – country report on mineral emissions. Contribution to 'The concerted action programme founded by the EU Environment Programme – AIR3'. Danish Institute of Agricultural and Fisheries Economics (1996).

Heissenhuber, A., Katzek, J., Meuzel, F., Ring, H. Agriculture and Environment. Economica Verlag, Bonn (1994).

Hellegers, P.J.G.J. Environmental policy for the control of nitrate pollution at farm level in the European Union, Conference proceedings of The Nitrate Directive and the agricultural sector in the European Union. The Hague, June 20–22, 1996.

MAFF. Code of Good Agricultural Practice for the protection of water. Ministry of Agriculture, Fisheries and Food, July 1991.

204

MAFF. Solving the nitrate problem; progress in research and development. Ministry of Agriculture, Fisheries and Food (1993).

MAAC. Canadian legislation, regulation and codes of practice for manure management. Ministry of Agriculture and Agri-Food Canada, Ottawa (1994).

Miller, M.H., Goss, M.J. Agricultural impacts on water quality: an Ontario perspective, National workshop proceedings on agricultural impacts on water quality. Canadian Agricultural Research Council, Ottawa (1992).

Ministry of the Environment, Denmark. Environmental impacts of nutrient emissions in Denmark (1991).

MLNV. Policy Document on Manure and Ammonia, Ministry of Agriculture, Nature Management and Fisheries, The Hague (1995).

MUBW. Nitrate in groundwater, Ministerium für Umwelt Baden-Wurtemberg, Stuttgart (1989).

MVW. Achtergrondnota Toekomst voor water, Ministerie van Verkeer en Waterstaat, Directoraat-Generaal Rijkswaterstaat, The Hague (1996).

MVW. Watersysteemverkenningen, Ministerie voor Verkeer en Waterstaat, Directoraat-Generaal Rijkswaterstaat, The Hague (1996).

Nagpal, N.K. Impact of agricultural practices on water quality: a British Columbia perspective, National workshop proceedings on agricultural impacts on water quality. Canadian Agricultural Research Council, Ottawa (1992).

OMAF. Best Management Practices 2, Livestock and Poultry Waste Management. Ontario Ministry of Agriculture and Food, Toronto (1993).

OSPAR. Nutrients in the convention area. Oslo and Paris Commissions, 1993.

OSPAR. Nutrients in the convention area. Oslo and Paris Commissions, 1995.

Paterson, B.A., Lindwall, C.W. Agricultural impacts on water quality in Alberta. National workshop proceedings on agricultural impacts on water quality. Canadian Agricultural Research Council, Ottawa (1992).

Patni, N.K. Regulatory aspects of animal manure utilisation in Canada. Proceedings 7th Technical Consultation on the ESCORENA Network on Animal Waste Management, Editor Hall, J.E. REUR Technical Series 34, Food and Agriculture Organisation of the United Nations, Rome (1994).

Pierzynski, G.M., Sims, J.T., Vance, G.F. Soils and environmental quality. Lewis Publishers, Boca Raton (1994).

Purser, J. Animal manure as fertiliser. Alaska Cooperative Extension, 100G-00340, Fairbanks Al (1994).

Racz, G. Effect of agriculture on water quality in Manitoba. National workshop proceedings on agricultural impacts on water quality. Canadian Agricultural Research Council, Ottawa (1992).

Richards, J., Milburn, P. The impact of agricultural activities on water quality in Atlantic Canada. National workshop proceedings on agricultural impacts on water quality, Canadian Agricultural Research Council, Ottawa (1992).

Reichow, K. Manure management planning guide for livestock operators. Minnesota Department of Agriculture St. Paul (1995).

Schepers, J.S., Moravek, M.G., Alberts, E.E. and Frank, K.D. Maize production impacts on groundwater quality. J Environ Qual 20, 12 (1991).

Schleef, K.H. and Kleinhanss, W. Mineral balances in agriculture in the EU, Part 1, the regional level. Institute of farm economics, Federal Agricultural Research Center (FAL), Braunschweig (1994).

Schleef, K.H. Impacts of the Nitrate Directive and the Fertilizer Decree on assessments at farm level for Germany. Conference Proceedings on The Nitrate Directive and the European Union, The Hague, June 20–22 (1996).

Schleef, K.H. and Kleinhanss, W. Regional Nitrogen balances in the EU Agriculture. Conference Proceedings on The Nitrate Directive and the Agriculture Sector in the EU, The Hague, June 20–22 (1996).

Schou, J.S. Implementation of nitrate policies in Denmark, in: Brouwer, F.M. and Kleinhanss, W. (Editors), Nitrate Standards in Europe. Vank Publishers (1997).

Sharpley, A.N., Hedley, M.J., Sibbesen, E., Hillbricht-Ilkowska, A., House, W.A., Ryszkowski, L. Phosphorus transfers from terrestrial systems to aquatic ecosystems. In: Tiessen, H. (editor), Phosphorus in the global environment, Scope 54, Wiley, Chichester (1995).

Smith, W. Water resources in a global environment: policy issues. National workshop proceedings on agricultural impacts on water quality. Canadian Agricultural Research Council, Ottawa (1992).

Somers, G. Agricultural impacts on water quality, a P.E.I. perspective. National workshop proceedings on agricultural impacts on water quality. Canadian Agricultural Research Council, Ottawa (1992).

USDA. Agriculture waste management field handbook. US Department of Agriculture, Soil Conservation Service, Washington DC (1992).

USEPA. Another look: national survey of pesticides in drinking water wells phase II report. US Environmental Protection Agency Office of Water, EPA 579/09–91–020, Washington DC (1992).

USEPA. Manual of Nitrogen Control, US Environmental Protection Agency Report EPA/625/R-93/010. Washington, DC (1993a).

USEPA. Guidance specifying management measures for sources of non-point pollution in coastal waters. US Environmental Protection Agency, Office of Water, Washington DC (1993b).

USEPA. Guide manual on NPDES regulations for concentrated animal feeding operations. US Environmental Protection Agency Office of Water Report EPA 833-B-95–001, Washington DC (1995).

van Eerdt, M. Mestproductie, mineralen uitscheiding en mineralen in de mest. Maandstatistiek Landbouw (CBS), 11 (1996).

van Gijseghem, D.E.L. Social correction to agro-environmental policies in Flanders. Proceedings of the conference on the Nitrate Directive and the agricultural sector in the European Union, The Hague, June 20–22 (1996).

Veenhuizen, M.A., Eckert, D.J., Elder, K., Johnson, J., Lyon, W.F., Mancl, K.M., Schnitkey, G. Ohio lifestock manure and waste water guide. Cooperative Extension Service, Ohio State University, Bulletin 604, Columbus Ohio (1992).

VLM. Jaarverslag Vlaamse Land Maatschappij Boekjaar 1994. Brussel (1995).

VLM. Gids bij het nieuwe mestdecreet. Brussel (1996).

Vosmer, M. Internationaal mest- en ammoniakbeleid; een literatuurstudie naar het mest- en ammoniakbeleid van België, Duitsland, Denemarken, Frankrijk, Groot-Brittainië, Nederland, Noorwegen, Zweden en Zwitserland. Stageverslag RIZA, Lelystad (1995).

ABOUT THE AUTHORS

Foppe de Walle is affiliate professor in the Department of Environmental Health at the University of Washington, Seattle, USA. He is also associated with and co-founder of ENERO, the European Network of Environmental Research Organisations, in which he represents TNO, the Netherlands Organisation for Applied Scientific Research in Delft. He is also consultant to KPMG-Environment in the Hague. He received his BSc and MSc from the State University in Wageningen in sanitary engineering and his PhD in environmental engineering from the University of Washington. He was also a faculty member at the University of Illinois in Urbana and at Stanford University, Menlo Park. His specialty is environmental technology, public health engineering and environmental policy. He has co-authored three books on environmental management for companies, public health in developing countries and the environmental protection of the Mediterranean. The author can be reached at P.O. Box 6040, 2600 JA Delft, Netherlands, by phone at 31-15-2696886, by fax at 31-15-2696887 and by e-mail at dewalsev@wxs.nl

Joke Sevenster is director of Promikron BV, a research management and consultancy firm in Delft. She received her BSc and MSc in sociology with a specialty in extension communication at the State University in Wageningen. She received her PhD in educational psychology at the University of Washington in Seattle, USA. She was an extension assistant at the University of Illinois in Urbana, and a researcher at the State University in Wageningen and the University of Limburg in Maastricht. She is the author of two books on extension education and public policy.

INDEX

208